D1619573

MIX
Papier aus verantwortungsvollen Quellen
Paper from responsible sources
FSC® C105338

Jiajun Zhu

Optical and electrical properties of topological insulator Bi_2Se_3

Anchor Academic
Publishing

Zhu, Jiajun: Optical and electrical properties of topological insulator Bi_2Se_3, Hamburg, Anchor Academic Publishing 2017

Buch-ISBN: 978-3-96067-160-2
PDF-eBook-ISBN: 978-3-96067-660-7
Druck/Herstellung: Anchor Academic Publishing, Hamburg, 2017

Bibliografische Information der Deutschen Nationalbibliothek:
Die Deutsche Nationalbibliothek verzeichnet diese Publikation in der Deutschen Nationalbibliografie; detaillierte bibliografische Daten sind im Internet über http://dnb.d-nb.de abrufbar.

Bibliographical Information of the German National Library:
The German National Library lists this publication in the German National Bibliography. Detailed bibliographic data can be found at: http://dnb.d-nb.de

Hermannstal 119k, 22119 Hamburg
http://www.diplomica-verlag.de, Hamburg 2017
Printed in Germany

Table of Contents

Preface

The Nobel Prize in physics 2016 was divided to David J. Thouless, F. Duncan M. Haldane, and J. Michael Kosterlitz for their theoretical discoveries of topological phase transitions and topological phases of matter. They opened the door on an unknown world where matter can assume strange states. Mathematical methods have been used to study these phases. Their pioneering work is useful for the possible application in materials science and electronics in the future.

Topological insulator is one of the hottest research topics in solid state physics. The current research topic of Dr. Jiajun Zhu is related to topological insulator Bi_2Se_3. As a DAAD fellowship holder, Dr. Jiajun Zhu is doing research in Helmholtz-Zentrum Dresden-Rossendorf (HZDR), Germany, since 2015. He works in the institute of ion beam physics and materials research with the institute director, Prof. Manfred Helm, and the head of spectroscopy department, Dr. Harald Schneider, whose PhD's supervisor was Klaus von Klitzing, Nobel Prize Laureate in 1985. After the quantum Hall states discovered in 1980 by Klaus von Klitzing, more and more topological states appeared, such as the quantum spin Hall effect, quantum anomalous Hall effect, topological insulators, Dirac semimetal and Weyl semimetal. Thousands of papers have been published to search, identify and manipulate these topological states, as it has unexpected fashion insights from many different fields, including semiconductor physics, materials science, spintronic, quantum field theory, topological field theory, and particle physics.

The first time Dr. Jiajun Zhu came into the region of physics electronics was 2004 in East China Normal University, where he got his PhD degree with a National Scholarship, awarded for top student national wide. Dr. Jiajun Zhu was a visiting scientist in Materials Department, University of Cambridge from 2012 to 2013. After that, Dr. Jiajun Zhu joined in Chinese Academy of Sciences as a research physicist & Sailing Grant holder. His research interest lies in the area of condensed matter and materials physics, as well as the energy applications of materials and devices.

This is the first book to describe the vibrational spectroscopies and electrical transport of topological insulator Bi_2Se_3, one of the most exciting areas of research in condensed matter physics. In particular, attempts have been made to summarize and develop the various theories and new experimental techniques developed over years from the studies of Raman scattering, infrared spectroscopy and electrical transport of topological insulator Bi_2Se_3. It is intended for material and physics researchers and graduate students doing research in the field of optical and electrical properties of topological insulators, providing them the physical understanding and mathematical tools needed to engage research in this quickly growing field. Some key topics in the emerging field of topological insulators are introduced.

There are many excellent books dedicated to the topological insulators and its related area. The book, Topological Insulators edited by Marcel Franz and Laurens Molenkamp, gives contemporary concepts of this topic. The book, Topological Insulators by Frank Ortmann, Stephan Roche, and Sergio O. Valenzuela, provides a full overview and in-depth knowledge about this hot topic, especially in angle-resolved photoemission spectrometry (ARPES), advanced solid-state Nuclear Magnetic Resonance (NMR) and scanning-tunnel microscopy (STM). The book, Topological Insulators: Dirac Equation in Condensed Matters by Shun-Qing Shen, presents a unified description of topological insulators from one to three dimensions based on the modified Dirac equation. The book, Topological Insulators and Topological superconductors by B. Bernevig, Andrei Hughes, and L. Taylor, comprehensively offers the concept of Berry phases Dirac fermions Hall conductance and zero modes on vortices in topological insulators and superconductors. The book, DFT Applications in Topological Insulators and Overdoped Cuprate: a study of topological insulator states of half-Heusler materials and the momentum density of overdoped LSCO by Wael Al-Sawai, uses DFT to study topological insulators and over doped high temperature cuprate superconductors. The book, Bulk and Boundary Invariants for Complex Topological Insulators by Emil Prodan and Hermann Schulz-Baldes, provides an overview of rigorous results on fermionic topological insulators.

It is apparent that the subjects related to topological insulators are very diverse and each book could merely describe a small portion of these topics. I apologize to those authors whose excellent works may have been overlooked among so many research papers in topological insulators.

This book is written by Jiajun Zhu supported by CSC-DAAD program on topological insulators, cooperated with Manfred Helm, Harald Schneider, Shengqiang Zhou, and Alexej Pashkin.

Acknowledgements

I would like to thank all the people who helped me during my topological insulator research work over the last a few years, especially for my colleagues in Helmholtz-Zentrum Dresden-Rossendorf, namely Manfred Helm, Harald Schneider, Shengqiang Zhou, Alexsiy Pashkin, Stephan Winnerl, Manos Dimakis, Uta Lucchesi, Leila Balaghi, Johannes Braun, Ivan Fotev, Jacob König-Otto, Denny Lang, Rakesh Rana, Faina Eßer, Johannes Schmidt, Ross Sheldon, Abhishek, Singh, Tina Tauchnitz, and Tommaso Venanzi.

Gunther Springholz in Institute of Semiconductor and Solid State Physics, Johannes Kepler University Linz provided high-quality topological insulator samples, which were very important to our research.

Senior scientists in Shanghai Institute of Technical Physics, namely Junhao Chu, Xiangjian Meng, Zhiming Huang, Guolin Yu, Jinglan Sun, Yun Hou, Shaowei Wang, and Professor Zhigao Hu in East China Normal University gave important guidance to applying for the overseas research program.

The research work was supported by Shanghai Sailing Program (Grant No. 15YF1413900), CSC-DAAD research scholarship (CSC No. 201500110019 and DAAD No. 57165010), the International Postdoctoral Exchange Fellowship Program (Grant No. 20150057) and Project funded by China Postdoctoral Science Foundation (Grant No. 2014M560357).

Jiajun Zhu

National Lab of Infrared Physics, Shanghai Institute of Technical Physics, Chinese Academy of Sciences;

DAAD fellow in Institute of Ion Beam Physics and Materials Research, Helmholtz-Zentrum Dresden-Rossendorf

June 2017

Abstract

This book introduces the recent results and progress of Raman spectra, electrical transport, and terahertz spectroscopy of topological insulator Bi_2Se_3, which is the topic of CSC-DAAD research scholarship program, held by Jiajun Zhu and his collaborators Manfred Helm, Harald Schneider, Shengqiang Zhou, and Alexej Pashkin in Helmholtz-Zentrum Dresden-Rossendorf. Our preliminary results have been published in Applied Physics Letters 109, 202103 (2016), with the title Lattice Vibrations and Electrical Transport in $(Bi_{1-x}In_x)_2Se_3$ Films.

Chapter 1 introduces the basic knowledge and recent progress of topological insulators, including topological phase transition, quantum Hall effect, and topology.

Chapter 2 is second generation material Bi_2Se_3, including crystal and band structure, substrate choice, and doping effect.

Chapter 3 is Raman spectroscopy, stokes and anti-stokes scattering, thickness dependence, Bi_2Se_3 films on GaAs and graphene, temperature and pressure dependence.

Chapter 4 is electrical transport, including weak antilocalization, linear magnetoresistance, Shubnikov-de Haas oscillations, Zeeman effect, influence of thickness and pressure, and doping effect.

Chapter 5 is infrared spectroscopy, including optical conductivity, plasmon and charge inhomogeneity, thickness dependent electronic properties, magneto optical spectroscopy, bandgap of Bi-based topological insulators.

Key words: Topological insulator Bi_2Se_3, Raman spectra, Transport, Terahertz

Chapter 1 Introduction

1.1 Phases of matter and topological phase transition

The Laureates of the Noble prize in physics 2016 opened a door on an unknown world where material exists in strange states.[1,2] They studied phenomena in a flat two-dimensional world, such as surfaces or inside extremely thin layers. This is different to the real world which is usually described by three dimensions, including length, width, and height. Correspondingly, the physics are different and new phenomena are continually discovered in these two dimensional world. Topological concept, a branch of mathematics, was used to interpret their discoveries in physics. When the temperature is low enough, close to absolute zero at -273 degrees Celsius, matter behaves in unexpected ways and new phases appear. Quantum physics becomes visible, as shown in Fig. 1.1.

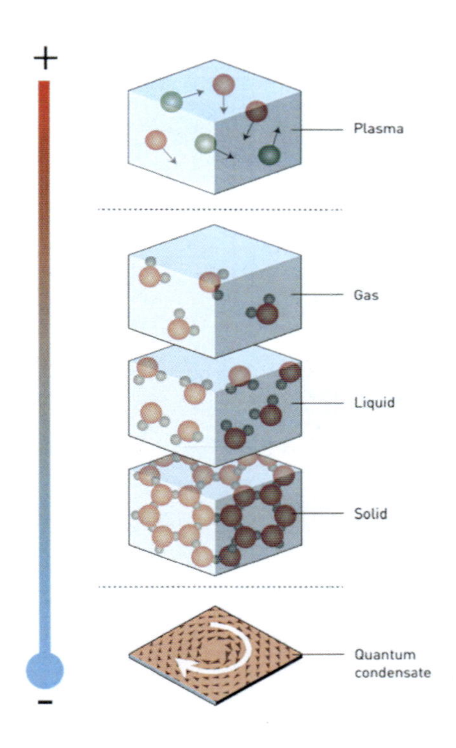

Figure 1.1 Phases of matter. Gas, liquid, and solid states are the most common phases of matter. In extremely high or low temperatures other states exist, from reference [1].

When temperature changes to a critical point, an ordinary phase transition occurs. For example, with increasing temperature up to above zero degree Celsius, ice melts into water. Topological phase transition is different from an ordinary phase transition. Small vortices in the flat material play an important role in a topological phase transition. Small vortices form tight pairs at low temperature, and a phase transition occurs when the temperature rises. The small vortices separate all of a sudden and sail off on their own. This was regarded as one of the most important condensed matter physics discoveries in 20th century. It is named as Kosterlitz-Thouless (KT) transition or Berezinskii-Kosterlitz-Thouless (BKT) transition. These three scientists challenged the then theory that thermal fluctuations destroy all order in matter in a flat 2D world including at absolute zero, and no ordered phases, no phase transitions. The BKT transition theory can be used for different types of materials in low dimensions, and it is confirmed by experiments.

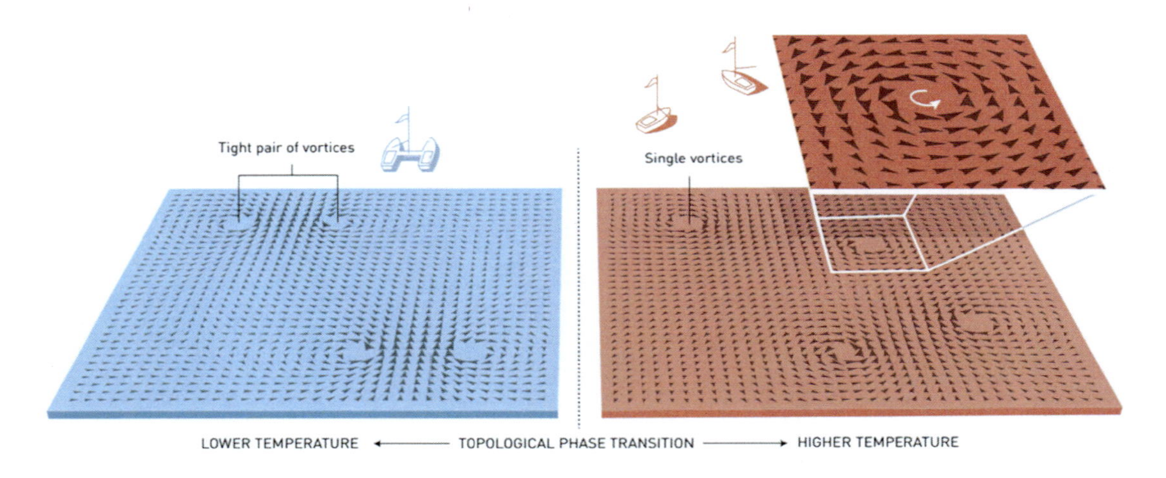

Figure 1.2 Topological phase transition in a thin layer of very cold matter described by Kosterlitz and Thouless from reference [1].

1.2 Quantum Hall effect

New states of matter are usually found by experimental developments and explained by new theory. Quantum mechanics was established and developed by Max Planck, Niels Bohr, Werner Heisenberg, Louis de Broglie, Arthur Compton, Albert Einstein, Erwin Schrödinger, Max Born, John von Neumann, Paul Dirac, Enrico Fermi, Wolfgang Pauli, Max von Laue, Freeman Dyson, David Hilbert, Wilhelm Wien, Satyendra Nath Bose, Arnold Sommerfeld, and others. And the quantum theory was considered to be well understood over several dozens of years.

Figure 1.3 The 1927 Solvay Conference in Brussels[2].

However, D. Thouless and D. Haldane proved that the previous theory was incomplete at low temperatures and in strong magnetic fields in 1983. That began the rapid developments to the theory of new phases of matter. D. Thouless described the quantum Hall effect using topology. Quantum Hall effect was discovered by German scientist Klaus von Klitzing, 1985 Nobel Prize Laureate. As claimed by Klaus von Klitzing, the quantum Hall effect was discovered at around 2 a.m., Feb. 5, 1980 at the High Magnetic Field Laboratory in Grenoble. [3] The experiments was measured in low temperature (4.2 K) to suppress disturbing scattering processes from electron-phonon interactions. A strong magnetic field at 19.8 Tesla was applied to get more information about microscopic details of the materials. It is a quantum mechanical version of the Hall effect, and the Hall conductance undergoes quantum Hall transitions to take on the quantized values. The plateau values in the Hall resistance are not influenced by the amount of localized electrons. It can be expressed by the equation $\rho_{xy}=h/ie^2$, where h is Planck constant, e is elementary charge and i is the number of fully occupied Landau levels.

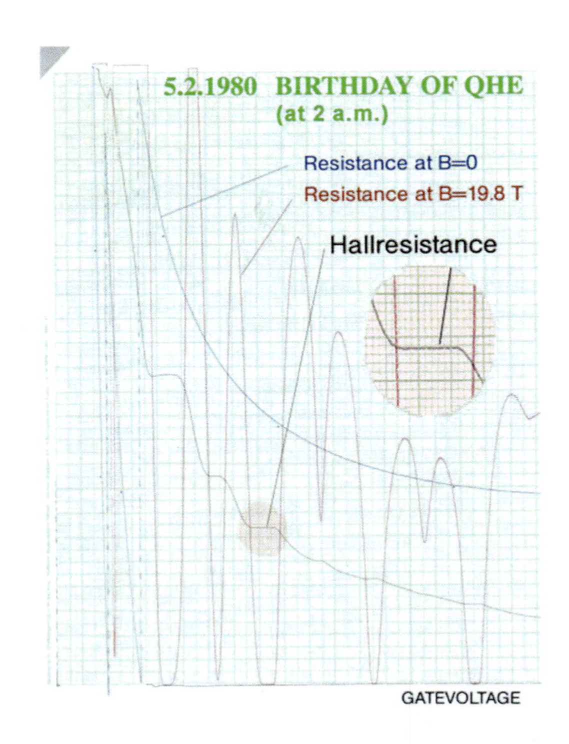

Figure 1.4 Experimental curve led to the discovery of quantum Hall effect from reference [3].

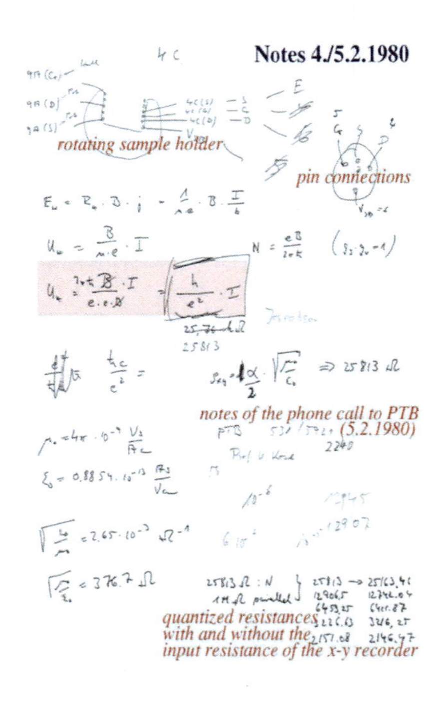

Figure 1.5 Copy of the original notes by Klaus von Klitzing. It led to the discovery of the quantum Hall effect from reference [3].

Resistance was measured at B=0 and 19.8 T at 4.2 K. Blue curve is the electrical resistance of the silicon field effect transistor as a function of the gate voltage. With increasing gate voltage, the electron concentration increases and the electrical resistance becomes gradually smaller and smaller. The Hall voltage is inversely proportional to the electron concentration, so the Hall voltage decreases with increasing gate voltage. Black curve is the Hall resistance, which can be calculated by Hall voltage divided by the current through the sample.

Quantum Hall effect is not easy to understand because the electrical conductance in the layer is merely particular values with extremely precise. When the temperature, magnetic field, or the doping level changes to some extent, the results are precisely the same. When it changes enough, the conductance of the layer vary only in integer steps. It becomes twice, triples, quadruples, and so on. At that time the physicists could not explain it, however, David Thouless used topology concept to interpret it.

1.3 Concept of topology

Topology is a math concept. Topologically, an orange and a bowl belong to the same category due to the smooth transforming between them. Apparently, a cup with a hole in the handle and a bowl do not belong to the same category. Topological objects can have one hole, two holes, three holes... but the hole number must be an integer. It is useful to describe the electrical conductance found in the quantum Hall effect, which merely changes in steps. The steps should be integer.

Figure 1.6 Topology is interested in properties of space that are preserved under continuous deformations, such as stretching and bending, but not tearing or gluing. Properties change step-wise, like the number of holes in between the orange and donut, or between the bowl and the cup. This Figure is *compiled by the author.*

1.4 Quantum Spin Hall effect

Integer quantum Hall effect was a tip of an iceberg. Later on, fractional quantum Hall effect (FQHE) was discovered in 1982. The 1998 Nobel Prize in Physics was awarded to Robert Laughlin, Horst Störmer, and Daniel Tsui for their discovery and explanation of the fractional Hall effect. Electron correlations lead to the appearance of fractionally charge quasiparticles, which are neither bosons nor fermions. They exhibit anionic statistics. The FQHE continues to be influential in theories about topological order, which do not have much relevance to topological insulators.

Spin Hall effect is a collection of relativistic spin-orbit coupling phenomena. [4] Electrical currents can generate transverse spin currents and vice versa in this phenomena. Spin Hall effect was discussed theoretically since 1971 [5] and it was confirmed experimentally in 2004.[6] Quantum spin Hall state exist in two-dimensional semiconductors with a quantized spin-Hall conductance and a vanishing charge-Hall conductance. It can be realized on lattice without the application of a large magnetic field. Kane and Mele proposed the first proposal for the existence of a quantum spin Hall state. [7] They developed a model for graphene by Haldane. [8] Their model contains a charge Hall conductance of zero but a spin-Hall conductance of $\boldsymbol{\sigma}_{xy}{}^{spin} = 2$. Quantum spin Hall state is non-trivial even after the introduction of spin-up spin-down scattering, [9] which destroy the quantum spin Hall effect. Quantum spin Hall state and topological insulator belong to different symmetry protected by topological states, which means they are different states of matter.

1.5 Topological insulator

Topological insulator is a new kind of insulator with a metallic boundary (Figure 1.7), originating from topological invariants. It cannot change as long as a material remains insulating. [11] The wave functions describe the electronic states, which span a Hilbert space. The Hilbert space has a nontrivial topology for topological insulators. The topological protection of the surface state is useful for low-power electronics and error-tolerant quantum computing. [12,13]

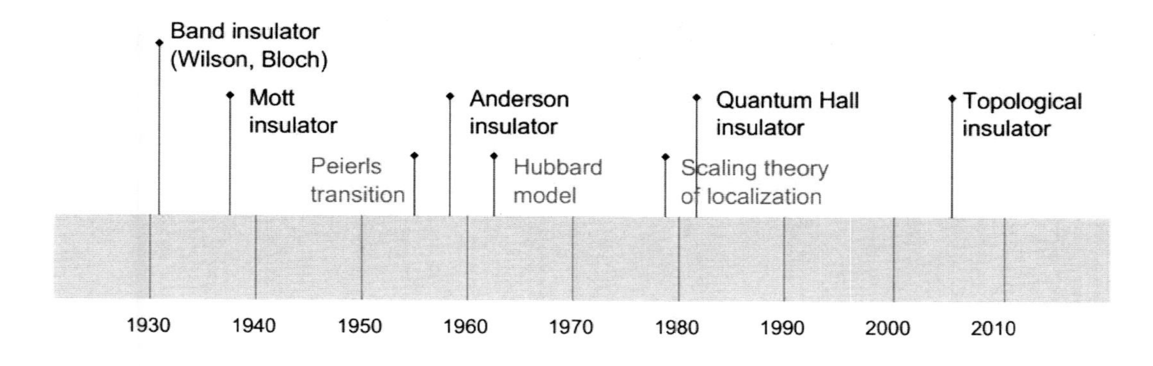

Figure 1.7 A brief history of insulators, from reference [10].

$Bi_{1-x}Sb_x$ is the first experimentally identified 3D topological insulator with a complicated surface structure and a small band gap. Therefore, scientists try to look for a large band gap topological insulator with a simple surface. Second generation topological insulators, especially Bi_2Se_3, Bi_2Te_3, and Sb_2Te_3 are theoretically predicted and experimentally observed with a single Dirac cone.

Bismuth is a pentavalent post-transition metal and its band structure features conduction and valence bands that overlap. This leads to pockets of holes close to the *T* point in the Brillouin zone as well as pockets of electrons close to the 3 equivalent *L* points. [14] At the *L* point, the valence is derived from antisymmetric (L_a) orbitals, while the conduction band is from symmetric (L_s) one. The energy gap is small. The band structure is changed by addition of antimony.

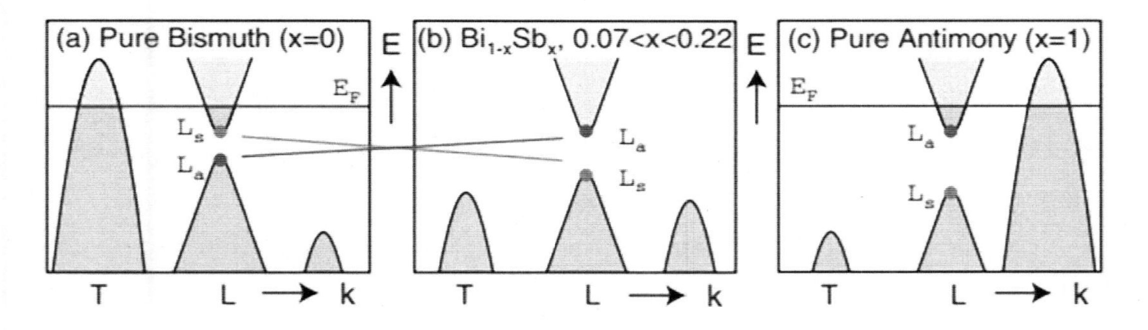

Figure 1.8 Schematic band structure of $Bi_{1-x}Sb_x$ from reference [14].

At around x=0.04, the gap closes and a truly massless three dimensional Dirac point can be realized. Further increasing Sb concentration reopens the gap with an inverted ordering. Above x=0.07, it shows insulating behavior, as the top of the valence band at T moves below the bottom of the conduction band at *L*. It becomes a direct gap insulator with a massive

Dirac-like bulk bands at Sb concentration of x=0.09, as the band at *T* drops below the valence band. Finally, for Sb concentration above 0.22, it restores the semimetallic state, as the valence band at a different point rises above the conduction band. [14] Both bismuth and antimony have a direct band gap, therefore, their valence bands can be topologically classified.

1.6 Angle-resolved photoemission spectroscopy

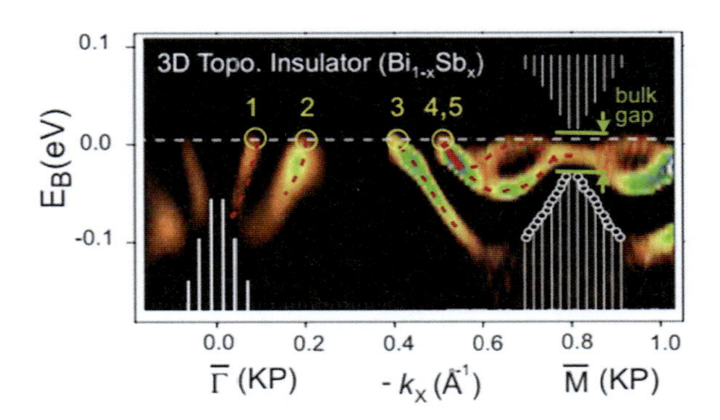

Figure 1.9 Topological surface states in $Bi_{1-x}Sb_x$ by Angle-resolved photoemission spectroscopy (ARPES) from reference [15].

It is a successful method to identify the two dimensional topological insulator by charge transport experiments. However, it is hard to be used in three dimensional materials. The main reason is the signal of the conductivity is much more subtle and the difficulty to separate the surface contribution and the bulk contribution. Angle-resolved photoemission spectroscopy (ARPES) is a useful technique to prob the topological character of the surface states. It can directly observe the distribution of the electrons in the reciprocal space of solid state materials. It uses a photon to eject an electron from a crystal. Then the surface/bulk electronic structure is determined by analyzing the momentum of the emitted electron.

Compared with transport measurements, ARPES can give the information on the distribution of spin orientations on the Fermi surface and estimate the Berry phase on the surface. [15-20] Figure 1.9 is the ARPES spectrum of $Bi_{0.09}Sb_{0.91}$. [21] Hsieh *et al.* observed the bulk energy bands and the *L* point, reflecting the nearly linear three-dimensional Dirac-like dispersion. The surface state structure of $Bi_{1-x}Sb_x$ is similar to those in Bi, [22] with an additional states near M point about 15 meV below E_F.

Reference

1. The Nobel Prize in Physics 2016 – Popular/Advanced Information. Nobelprize.org. Nobel Media AB 2014. Web. 4 Dec. 2016.

2. This is a file from the Wikimedia Commons, which is a freely licensed media file repository. Photograph by Benjamin Couprie, Institut International de Physique Solvay, Brussels, Belgium.

3. Klaus von Klitzing, 25 years of quantum Hall effect (QHE), A personal view on the discovery, physics and applications of this quantum effect, Séminaire Poincaré 2, 1 (2004).

4. Jairo Sinova, Sergio O. Valenzuela, J. Wunderlich, C. H. Back, and T. Jungwirth, Spin Hall effects, Reviews of Modern Physics, 87, 1213 (2015).

5. M. I. D'yakonov and V. I. Perel, Possibility of orienting electron spins with current, JETP Letters 13, 467 (1971).

6. Y. K. Kato, R. C. Myers, A. C. Gossard, and D. D. Awschalom, Observation of the Spin Hall Effect in Semiconductors, Science 306, 1910 (2004).

7. C. L. Kane and E. J. Mele, Quantum Spin Hall Effect in Graphene, Physical Review Letters 95, 226801 (2005).

8. F. D. M. Haldane, Model for a Quantum Hall Effect without Landau Levels: Condensed-Matter Realization of the "Parity Anomaly", Physical Review Letters 61, 2015 (1988).

9. C. L. Kane and E. J. Mele, Z2 Topological Order and the Quantum Spin Hall Effect, Physical Review Letters 95, 146802 (2005).

10. Ming Che Chang, Basics of topological insulator, Talk in Nanyang Technological University, 18 Nov. 2011.

11. Joel E. Moore, The birth of topological insulators, Nature, 464, 194 (2010).

12. I. Žutić, J. Fabian, and S. Das Sarma, Spintronics: Fundamentals and applications, Reviews of Modern Physics 76, 323 (2004).

13. Chetan Nayak, Steven H. Simon, Ady Stern, Michael Freedman, Sankar Das Sarma, Non-Abelian Anyons and Topological Quantum Computation, Reviews of Modern Physics 80, 1083 (2008).

14. M. Z. Hasan and C. L. Kane, Colloquium: Topological Insulators, Reviews of Modern Physics 82, 3045 (2010).

15. Miguel M. Ugeda, Aaron J. Bradley, Yi Zhang, Seita Onishi, Yi Chen, Wei Ruan, Claudia Ojeda-Aristizabal, Hyejin Ryu, Mark T. Edmonds, Hsin-Zon Tsai, Alexander Riss, Sung-Kwan Mo, Dunghai Lee, Alex Zettl, Zahid Hussain, Zhi-Xun Shen, and Michael F. Crommie, Characterization of collective ground states in single-layer $NbSe_2$, Nature Physics, 12, 92 (2016).

16. M. Xia, J. Jiang, Z. R. Ye, Y. H. Wang, Y. Zhang, S. D. Chen, X. H. Niu, D. F. Xu, F. Chen, X. H. Chen, B. P. Xie, T. Zhang, and D. L. Feng, Angle-resolved Photoemission Spectroscopy Study on the Surface States of the Correlated Topological Insulator YbB_6, Scientific Reports 4, 5999 (2014).

17. Chang-Jong Kang, J. D. Denlinger, J. W. Allen, Chul-Hee Min, F. Reinert, B. Y. Kang, B. K. Cho, J.-S. Kang, J. H. Shim, and B. I. Min, Electronic Structure of YbB_6: Is it a Topological Insulator or Not? Physical Review Letters 116, 116401 (2016).

18. S.V. Ramankutty, N. de Jong, Y.K. Huang, B. Zwartsenberg, F. Massee, T.V. Bay, M.S. Golden, E. Frantzeskakis, Comparative study of rare earth hexaborides using high resolution angle-resolved photoemission, Journal of Electron Spectroscopy and Related Phenomena 208, 43 (2016).

19. E. Frantzeskakis, N. de Jong, J. X. Zhang, X. Zhang, Z. Li, C. L. Liang, Y. Wang, A. Varykhalov, Y. K. Huang, and M. S. Golden, Insights from angle-resolved photoemission spectroscopy on the metallic states of $YbB_6(001)$: $E(k)$ dispersion, temporal changes, and spatial variation, Physical Review B 90, 235116 (2014).

20. Chris Jozwiak, Jonathan A. Sobota, Kenneth Gotlieb, Alexander F. Kemper, Costel R. Rotundu, Robert J. Birgeneau, Zahid Hussain, Dung-Hai Lee, Zhi-Xun Shen, and Alessandra Lanzara, Spin-polarized surface resonances accompanying topological surface state formation, Nature Communications 7, 13143 (2016).

21. D. Hsieh, D. Qian, L. Wray, Y. Xia, Y. S. Hor, R. J. Cava, and M. Z. Hasan, A topological Dirac insulator in a quantum spin Hall phase, Nature 452, 970 (2008).

22. F. Patthey, W. D. Schneider, and H. Micklitz, Photoemission study of the Bi(111) surface, Physical Review B 49, 11293 (1994).

Chapter 2 Second generation material Bi_2Se_3

2.1 Crystal and band structure

$Bi_{1-x}Sb_x$ has a high two dimensional carrier mobility of ~10^4 cm^2/Vs, which make it easy to study two dimensional quantum transport.[1] However, it has a complicated surface structure and its band gap is small. Therefore, scientists try to find topological insulators with a larger band gap and a simple surface structure. Bi_2Se_3, Bi_2Te_3, and Sb_2Te_3 crystallize in tetradymite structure. They crystallize in the rhombohderal crystal structure with the space group D^5_{3d} with five atoms in a unit cell. It consists of covalently bonded quintuple layers, as shown in Figure 2.1. Van der Waals force weakly interacts between the quintuple layers.

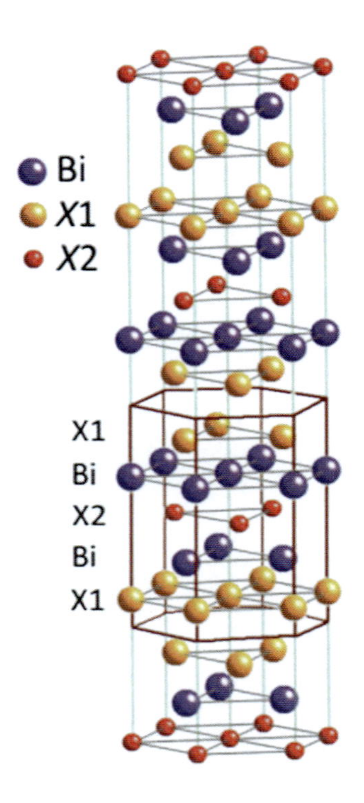

Figure 2.1 Crystal structure of Bi_2Se_3 (X1=X2=Se), Bi_2Te_3 (X1=X2=Te), and Bi_2Te_2Se (X1=Te, X2=Se).[2] The brown cage shows a quintuple layer (QL) which is roughly 1 nm thickness. Each QL layer consists of five atomic layers, stacking as X1-Bi-X2-Bi-X1 along the trigonal axis. Ionic covalent bonds exist between neighboring Bi-X layers.

The surface state structure of Bi_2Se_3 shows an almost idealized Dirac cone with only slight curvature in Figure 2.2(a), while that of Bi_2Te_3 is a little bit complicated. The Dirac point of Bi_2Te_3 is under the top of the valence band, as shown in Figure 2.2(b). This makes it impossible to identify the surface transport properties from the bulk carriers. On the other hand, the constant energy contour of the Dirac cone for Bi_2Se_3 is nearly spherical, while it is apparently hexagonal warping for Bi_2Te_3, as shown in Figures 2.2(c) and 2.2(d), respectively. A k^3 term leads to the warping. This term originates from cubic Dresselhaus spin-orbit coupling at the surface of rhombohedral structure.[3] The hexagonal warping causes strong quasiparticle interference [4] and a finite spin polarization.[5]

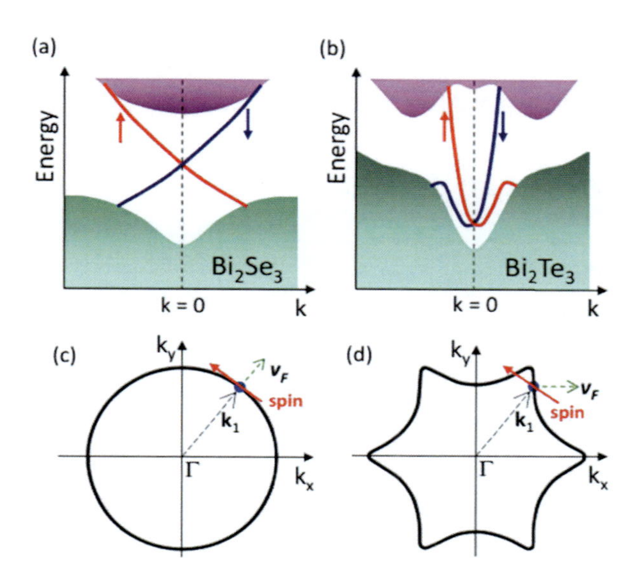

Figure 2.2 Band structures comparison between (a) Bi_2Se_3 and (b) Bi_2Te_3.[6] The surface states are spin non-degenerate and helically spin polarized. (c) Representative constant energy contours of the Dirac cones for Bi_2Se_3 and (d) Bi_2Te_3. In Bi_2Se_3 and Bi_2Te_3, the spin vector is perpendicular to the wave vector, while the Fermi velocity vector can be non-orthogonal to the spin vector in Bi_2Te_3 caused by the hexagonal warping. This leads to strong quasiparticle interference.

Due to naturally occurring crystalline defects, Bi_2Se_3 and Bi_2Te_3 are degenerately doped. This means their transport properties are dominated by bulk carriers. However, the surface state of Bi_2Te_2Se with a tetradymite structure shown in Figure 2.1, is metallic *n*-type. As a matter of fact, Bi_2Te_2Se is the first three dimensional topological materials to own a reasonably bulk-insulating behavior. The surface contribution of Bi_2Te_2Se is about 6%, [2] while it is merely 0.3% in Bi_2Te_3. [7]

2.2 Substrate choice

Topological insulators can be grew by chemical vapor deposition, [8,9] bulk Bridgman growth, [10] wet chemical synthesis, [11] and molecular beam epitaxy (MBE). [12,13] The MBE technique can keep accurate film thickness with very good doping control, and potential integration of heterostructures for device structures. The MBE growth of topological materials mechanism is different from that of conventional materials which require a strict lattice-matching. Van der Waals bonds between the QLs relax the lattice-matching condition. Therefore, many substrates have been selected for the growth of topological materials, as shown in Table 2.1.

Substrate	Bi_2Se_3	Sb_2Te_3	Bi_2Te_3
Graphene	-40.6%	-42.3%	-43.8%
Si	-7.3%	-9.7%	-12.3%
CaF_2	-6.8%	-9.2%	-11.9%
GaAs	-3.4%	-5.9%	-8.7%
CdS	-0.2%	-2.8%	-5.7%
InP	0.2%	-2.3%	-5.3%
BaF_2	5.9%	2.8%	0.1%
CdTe	10.7%	7.8%	4.6%
Al_2O_3	14.9%	12%	8.7%
SiO_2	18.6%	15.5%	12.1%

Table 2.1 Lattice mismatch of different substrates.[14]

During the growth, reflection high energy electron diffraction (RHEED) is commonly used. The distance between the two first-order stripes are the d-spacing. It is inversely proportional to the lattice constant. The lattice is nearly fully relaxed to the topological films, which can be known from the *d*-spacing reaching a constant value.

2.3 Doping effect

2.3.1 Ca doping to control Fermi level

The bulk carrier density of topological insulators is relatively high due to the defects. This suppresses the surface states conduction. The most commonly observed defects in Bi_2Se_3 are the vacancy of Se atoms. In order to lower the bulk carrier density as well as to control the Fermi level into the bulk bandgap, counter doping method is widely used. Ca is a *p*-type dopant and it replaces Bi in Bi_2Se_3.

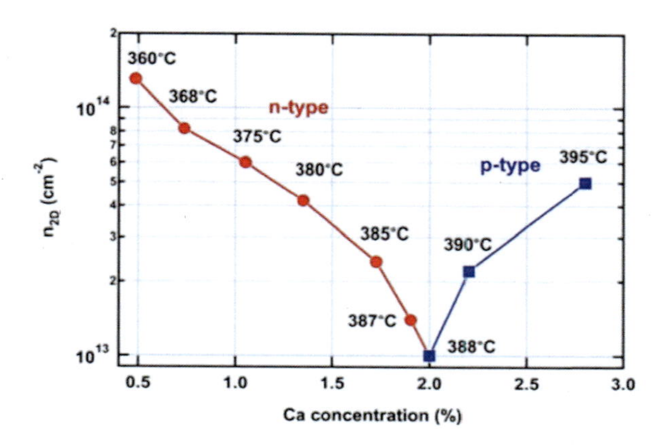

Figure 2.3 two dimensional carrier density of $Ca_xBi_{2-x}Se_3$. A systematic change of carrier density and type is shown. On top of the data is the temperature of Ca cell. Note that the y-axis is in Logarithmic coordinates.[15]

Figure 2.3 shows a systematic study of Ca doped Bi_2Se_3 using a Bridgman growth technique. The Fermi level moves continually across the Dirac point, with increasing Ca concentration. It changes from *n*-type material to *p*-type one. The lowest carrier density of Ca concentration at 2% is ten times smaller than the pure sample. The Ca doping concentration is tuned by adjusting the cell temperature of Ca in MBE growth. MBE-grown Bi_2Te_3 is usually n-type while Sb_2Te_3 is p-type. Therefore, the use of $(Bi_xSb_{1-x})_2Te_3$ alloy may tend to be a useful way of counter doping.

As a widely studied topological insulator, it is difficult to achieve low bulk conductivity for Bi_2Se_3, which depend on the Se partial pressure in synthesis. Although Bi_2Te_3 can be made n or p type through variation in the Bi:Te ratio, reported transport studies of Bi_2Se_3 shows p-type behavior. Native Bi_2Se_3 is usually n-type due to Se vacancies, which act as electron donors. The chemical properties of Bi and Te are similar, which leads to antisite defects as the primary source of carrier doping in binary compounds. However, tendency for Bi and Se mixing in Bi_2Se_3 is little. The main structural defect leading to electron doping is doubly charged Se vacancies.[16] Compensation doping is a common way to counteract the presence of Se vacancies. It requires careful control over p-type dopant concentrations, but it will result high bulk conductivity. The use of Ca substitution for Bi creates a negatively charged defect, which generates holes to compensate the electrons created by the Se vacancies.

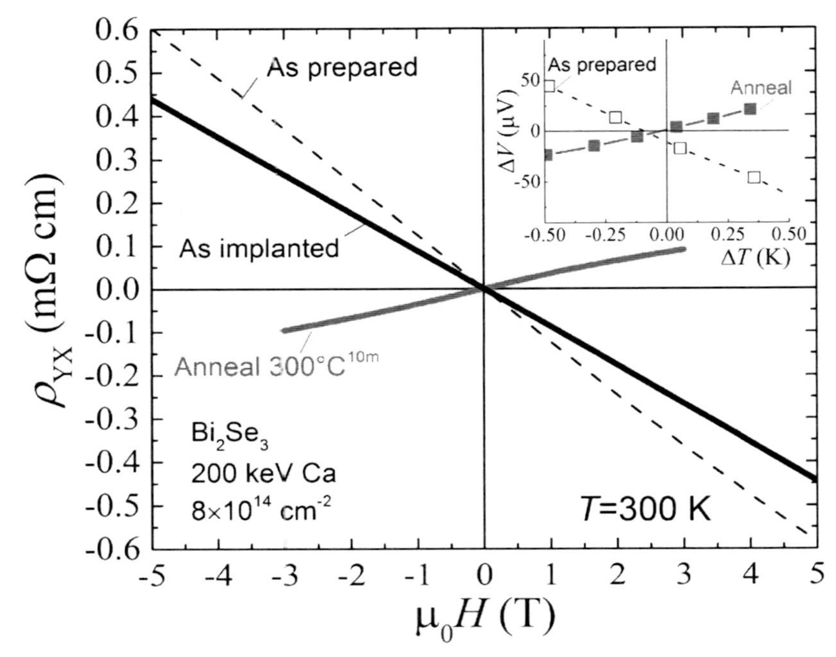

Figure 2.4 Hall resistivity versus magnetic field for as prepared Bi_2Se_3 films, Ca implanted films and the same films after annealing. Note that after annealing, the films are p-type. Inset is Seebeck voltage ΔV as a function of temperature gradient ΔT.[17]

This is particular important as Sharma et al. demonstrated that ion implantation is a useful way for the p-type doing of Bi_2Se_3 with a precise control over the number and spatial distribution of dopants.[17] Ion beam modification has two advantages compared with other approaches. First of all, it enables further control over carrier concentration to achieve low bulk conductivity. Secondly, it is efficient to fabricate inhomogeneous planar device structures. After ion implantation, the structure of material is usually damaged but such issue can be solved by a low temperature annealing. In order to mitigate ion beam damage and prevent Se loss during annealing, a thin layer of Al_2O_3 is coated. Calcium ion implantation broadens peak in X-ray diffraction and proper annealing process reduces the peak broadening. The annealing temperature is 350 °C, which is far below the melting temperature 700 °C. Therefore, recrystallization does not happen. It reflects that annealing removes ion beam damage.

Figure 2.4 shows room-temperature Hall measurements on as prepared films, films after implantation, and annealed films with implantation. It is clear that as both prepared films and implanted films are n-type, while annealed implanted films is p-type. The slope of Hall resistivity ρ_{YX} for the as implanted films is smaller than that for the as prepared one, indicates a higher carrier concentration due to the more Se vacancies from the introduction of ion beam modification. Note that Calcium is the simplest p-type dopant in Bi_2Se_3. The p-type behavior was reported to be induced through low-level substitutions (1% or less) of Ca for Bi. Fermi level is lowered into the valence band by about 400 meV in 2% Ca doped Bi_2Se_3 compared to the n-type material.[18] Moreover, nanoclusters of CaSe are found in Ca doped Bi_2Se_3 films.[19] The interface between insulator CaSe and topological insulator Bi_2Se_3 is topological nontrivial interface state.

2.3.2 Cr doping for ferromagnetism

Topological insulator has a conducting linearly dispersed Dirac surface state as well as a bulk energy gap. The surface state is predicted to become massive if time reversal symmetry is broken, and to become insulating if the Fermi energy is positioned inside both the surface and bulk gaps. Magnetic dopant Cr is a kind of transition metal, which is introduced into the Bi_2Se_3 to break the time reversal symmetry. Ferromagnetic order forms by introducing magnetic impurities into a topological insulator host lattice, as suggested by theoretical calculations. [20] Thus, the surface band will open a gap.

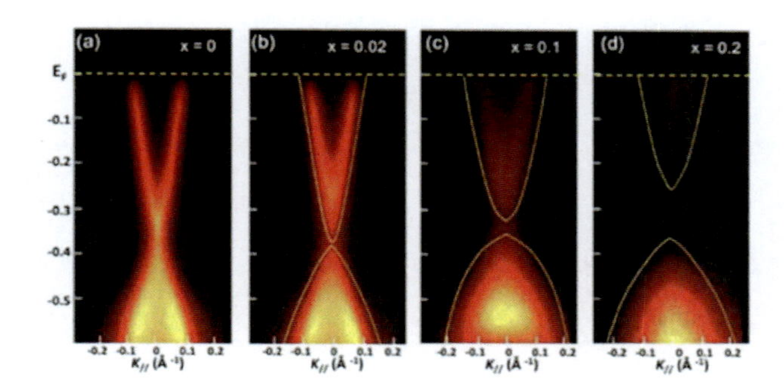

Figure 2.4 ARPES data of $Bi_{2-x}Cr_xSe_3$ films on Si with the thickness of 50 QL. With increase of Cr concentration, the surface state deviates from the original linear massless Dirac fermion state and it becomes broader and broader. Then a larger bandgap on the surface is open. The data were taken at 10 K.[20]

The bulk ferromagnetic property can be confirmed by magnetic and transport measurements. Below the Curie temperature, the Hall resistance R_{xy} becomes nonlinear due to the anomalous Hall effect. The magnetic direction is out of plane, parallel to the z-direction. The direction is important because the revelation of the anomalous Hall effect needs a thin ferromagnetic layer with the out-of-plane magnetism.

2.3.3 Cu doping for superconductivity

A superconductor can be regarded as topological superconductor if the topological number of the occupied states is nonzero.[21-29] The existence of gapless boundary states is its main characteristic feature. When it has a boundary and a topologically trivial state, it arises a mismatch of topology. This cannot be resolved without a singularity near the boundary. Such singularity is realized as gapless boundary states.[29] $Cu_xBi_2Se_3$ is the first topological insulator showing superconductivity with Curie temperature up to about 3.8 K.[30]

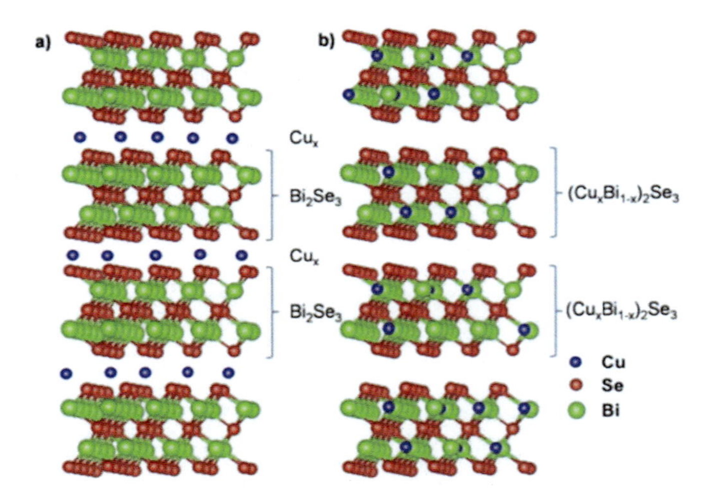

Figure 2.5 Schematic structure of $Cu_xBi_2Se_3$.[14] (a) In this case Cu^+ intercalated into Van der Waals gaps. (b) In this case Cu replaces Bi atoms. Cu dopant has two roles. On the one hand, Cu acts as an intercalant in the Van der Waals gaps, as shown in Figure 2.5(a). On the other hand, Cu replaces Bi, as shown in Figure 2.5(b).

The formation of substitutional Cu in replacing the Bi site cannot be totally avoided when the Cu^+ ion intercalates into the van der Waals gaps. Cu concentration below 15% is never superconducting for single crystals. It is still difficult to achieve the intercalation of Cu^+ only for bulks, but this problem can be solved by MBE growth, which makes it an interesting area to pursue.

2.4 A broad tenability for $(Bi_{1-x}In_x)_2Se_3$

2.4.1 Transport properties

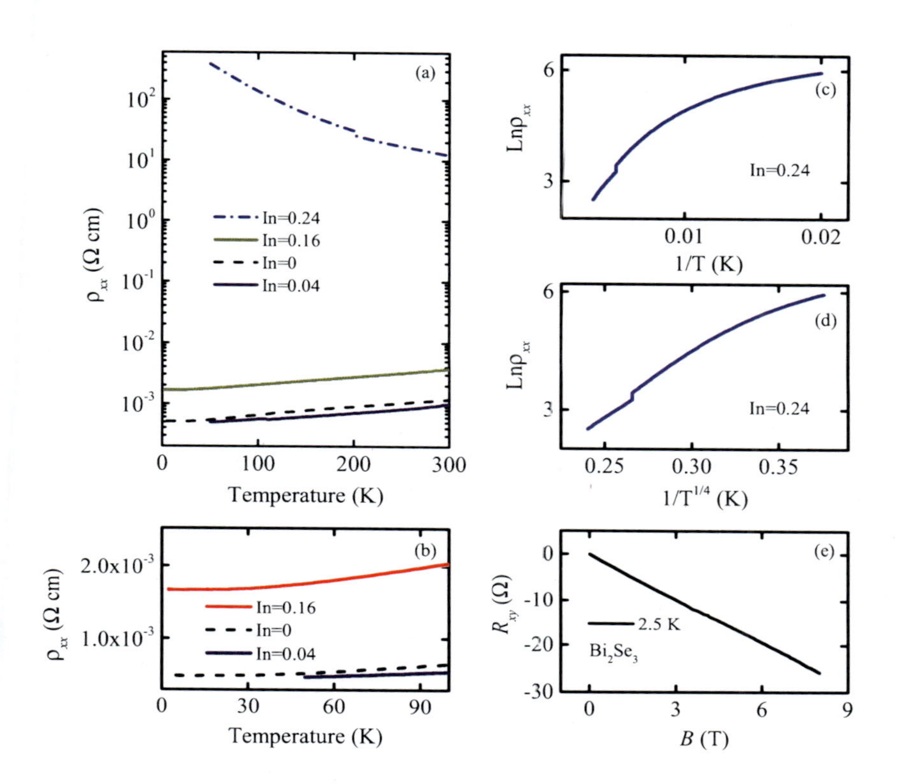

Figure 2.6 (a) Resistivity of $(Bi_{1-x}In_x)_2Se_3$ films below 300 K and (b) in the low temperature region. Below x=16%, it is metallic, as their resistivity decreases with decrease of temperature. (c) and (d) the resistance is plotted vs 1/T and $1/T^{1/4}$, which looks less nonlinear than the case of 1/T. This means variable-range hopping model $R(T)=R_0exp[(T_0/T)^{1/4}]$ fits better than thermal activation model $R(T)=R_0exp(E_a/k_BT)$. (e) Hall resistance versus magnetic field.[32] The sign of Hall resistance shows n-type conduction.

Bi_2Se_3 is a proto-typical topological insulator with a large nontrivial bulk gap of about 0.3 eV and a simple single Dirac cone surface state at the Γ point of the Brillouin zone, while In_2Se_3 is a topologically trivial band insulator with a band gap of 1.3 eV. Since the bulk Fermi energy E_F of Bi_2Se_3 stays in the conduction band due to intrinsic Se vacancies, it can be rather regarded as a topological metal. $(Bi_{1-x}In_x)_2Se_3$ shows a broad tunability from a topological-metal state to a truly insulating state through three quantum phase transitions by transport measurements.[31,32] Brahlek *et al.* found that a phase transition from a topologically nontrivial metal to a trivial metal occurs around indium concentration x=6%.

Above x=15% the metal becomes a variable-range hopping-insulator. Above x=25%, it becomes a true band insulator.[31] This was convinced by Zhu *et al.* [32] and Lou *et al.* [33] with some difference in indium contents.

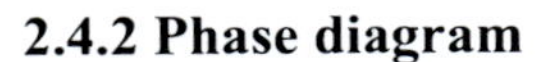

2.4.2 Phase diagram

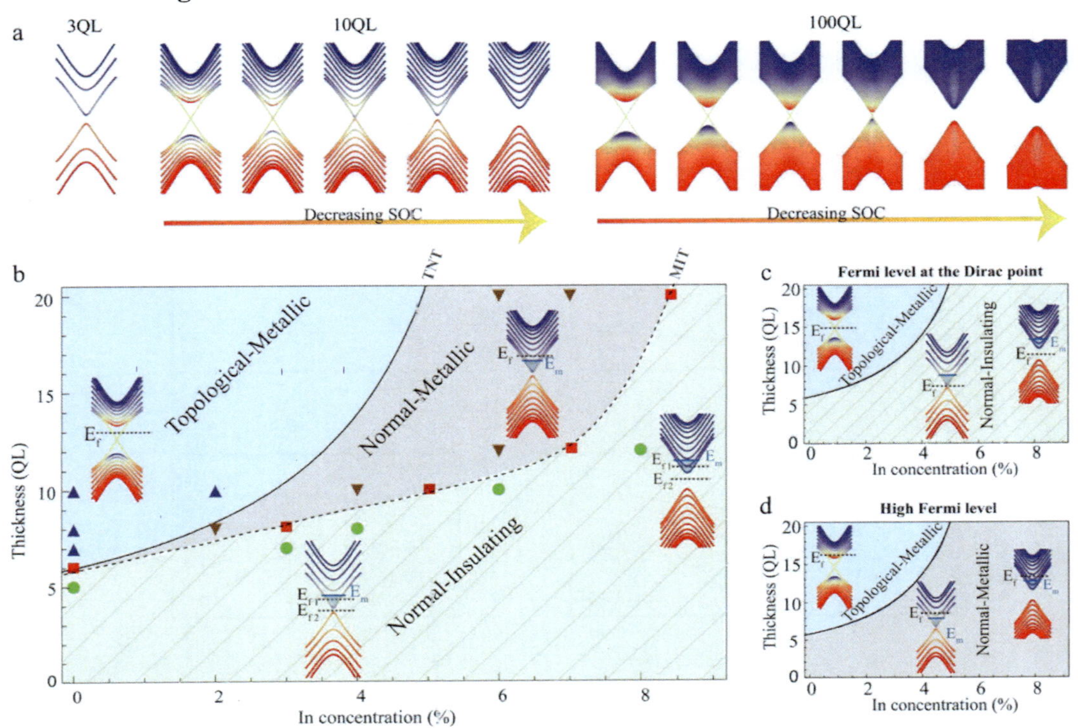

Figure 2.7 (a) Schematic of the topological phase transition process for 3 QL, 10 QL, and 100 QL thicknesses with spin-orbit coupling (SOC) weakening. (b) Phase diagram of indium doped Bi_2Se_3. Blue triangles is topological metallic data with $k_Fl>1$; green circles are normal insulating data where $k_Fl<1$; brown triangles are normal metallic data with $k_Fl>1$; red squares show data with $k_Fl=1$. Here $k_F=(3\pi^2 n_{2d}/t)^{1/3}$ is the three dimensional Fermi wave vector, $l=(2\pi h\mu/e)(3\pi^2 n_{2d}/t)^{1/3}$ is the mean free path, t is the film thickness, n_{2d} is the total sheet carrier density, and μ is the mobility. According to the concept of Ioffe-Regel rule, $k_Fl>1$ means a material is metallic, and metal insulator transition occurs at $k_Fl=1$.[34] In the region of normal insulating, E_{F1} and E_{F2} are two possible locations of the surface Fermi level. E_{F1} is below the mobility edge E_m while E_{F2} is inside the surface hybridization gap. (c) Phase diagram of the ideal case when the Fermi level is at the Dirac point. (d) Phase diagram of the high Fermi level above the bottom of the conduction band, which corresponds to $(Bi_{1-x}In_x)_2Se_3$ from the results of Brahlek et al..[31] Reprinted with permission from ACS.[35]

2.4.3 Terahertz range conductance

Figure 2.8 shows the transmittance spectra of $(Bi_{1-x}In_x)_2Se_3$ films. The data are fitted by a Drude-Lorentz model. The high-frequency dielectric constant is fixed to 30 due to screening by the valence electrons, as estimated from single crystal optical measurements.[36] Indium doping tends to increase the phonon frequency due to the lighter indium mass and to increase Drude scattering rate, which can be explained by increasing impurity scattering. The DC electrical and optical conductivities of the pure sample are slightly smaller than for the sample with indium concentration of 4%, which might be due to a slight variation of crystalline quality. The Drude term of x=24% sample vanishes, indicating it is not metallic. The peak is assigned to the E_u mode because the A_u mode is polarized normal to the layers and cannot be excited under the normal incidence. The E_u mode can be ascertained by polarizing the incident light perpendicularly to the c-axis.

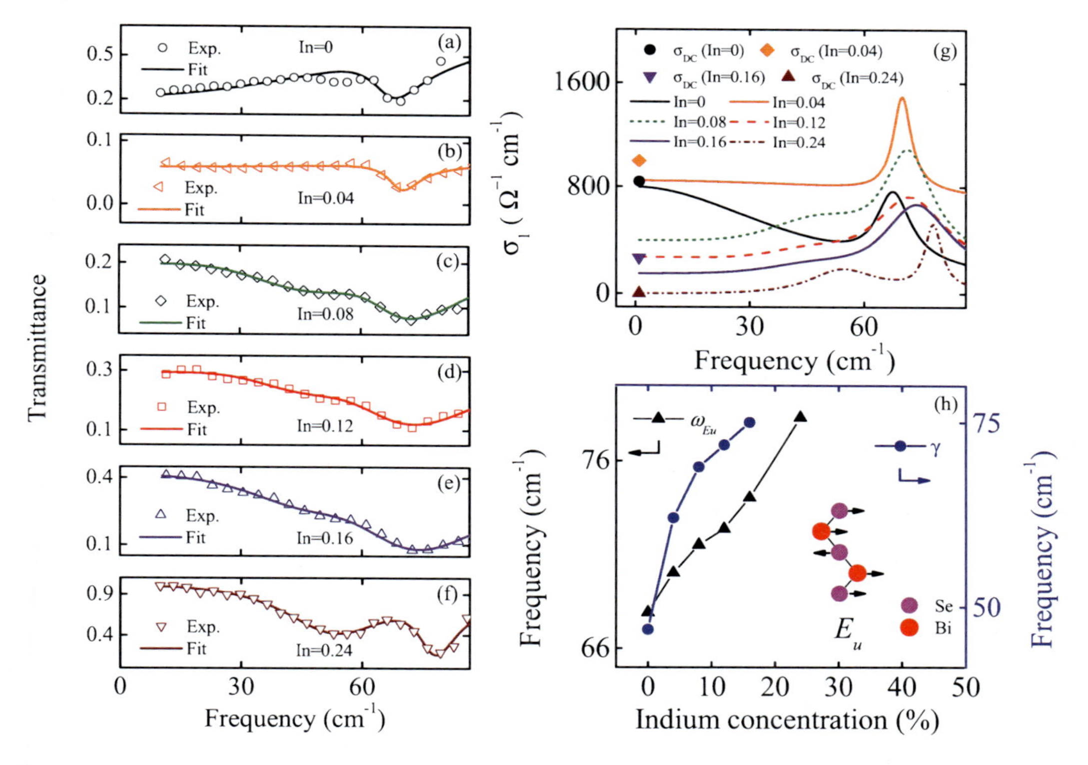

Figure 2.8 (a)-(f) Room temperature transmittance spectra of $(Bi_{1-x}In_x)_2Se_3$ films on BaF_2 (111) substrates. (g) Real part of the optical conductivity and its comparison with electrical conductivity from the resistivity measurements, which are shown in Figure 2.6. Phonon mode E_u and Drude scattering rate γ as a function of indium concentration.[32]

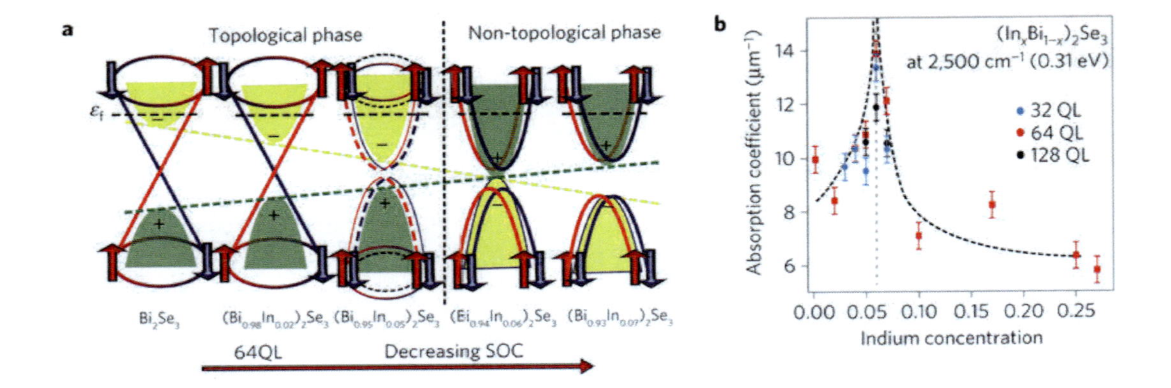

Figure 2.9 (a) A schematic of bulk band inversion. Valence band is indicated by the solid green color and plus sign, while conduction band is by yellow color and minus sign. (b) Infrared absorption at 2500 cm^{-1} (0.31 eV), which is close to the bulk bandgap of Bi_2Se_3 around 350 meV. The curved dashed line is a guide to the eye.[37]

Wu et al. reported a sudden collapse in the transport lifetime that indicates the destruction of the topological phase using time-domain terahertz spectroscopy and optical absorption measurements in the mid-infrared range.[37] The mid-infrared absorption coefficient at 0.31 eV has an obvious peak near indium concentration at x=6% for different film thicknesses, as shown in Figure 2.9. A closing of the bandgap increase the absorption because the low energy joint density of states increases. Therefore, the peak in the mid-infrared range absorption with indium concentration at 6% is consistent with a topological phase transition, related to the closing and reopening of the bandgap.

2.4.4 Nonlinear terahertz response

Dirac electrons in topological insulators have linear energy dispersion, a high mobility, a tunable density, a protection against backscattering through the spin-momentum locking mechanism, and a obvious nonlinear optical behavior. Study of nonlinear optical response is important to realize fundamental applications including coherent control of excitations in condensed matter. Giorgianni et al. demonstrated an electromagnetic-induced transparency of Bi_2Se_3 under a terahertz electric field.[38] Their discovery as well as harmonic generation and charge mobility reduction opens the path to tunable terahertz nonlinear optical TI-based devices.

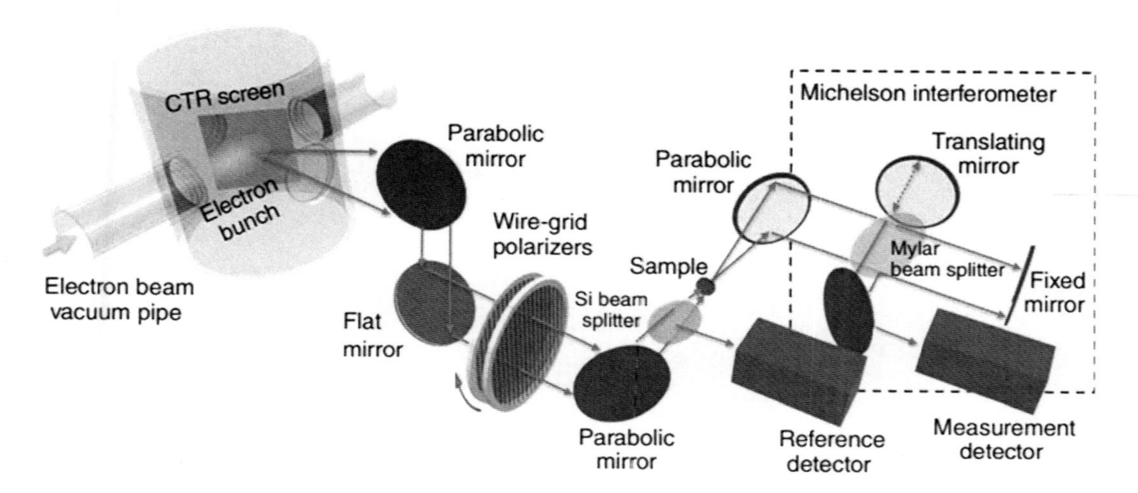

Figure 2.10 Experimental set-up.[38] Terahertz pulses are produced at SPARC_LAB, which can deliver broadband THz pulses with femtosecond shaping with the linac driven coherent THz radiation source as well as generate an adjustable train of electron bunches with a sub-picosecond length and with sub-picosecond spacing.[39] Terahertz source is the emission of coherent transition radiation.[40]

Blue arrows indicate terahertz radiation. Terahertz radiation is reflected at 90 degree with respect to the electron beam direction. It transmits through a quartz window to an off-axis parabolic mirror, which produces collimated radiation. Film transmittance is normalized to that of the substrate, so the substrate response is independent of the amplitude of the terahertz field. Therefore, the nonlinear terahertz effect is merely due to the Bi_2Se_3 films.

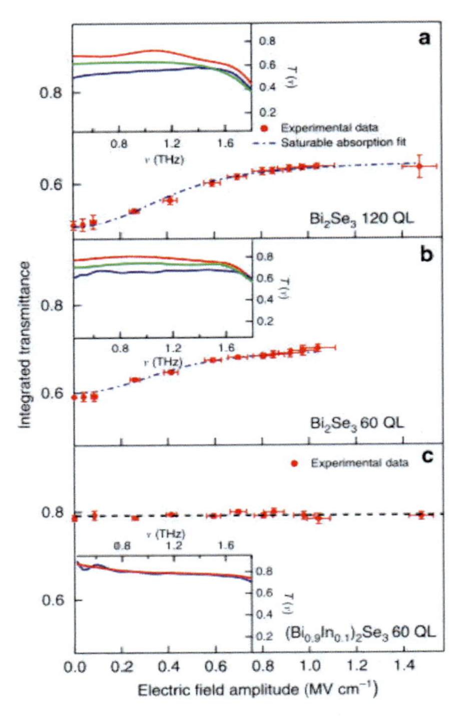

Figure 2.11 Terahertz nonlinear behavior of Bi_2Se_3 films.[38] Integrated transmittance of Bi_2Se_3 films with (a) 120 QL, (b) 60 QL, and (c) $(Bi_{0.9}In_{0.1})_2Se_3$ films versus the incident terahertz electric field amplitude. Red dots are experimental data. Fitting data with a saturable absorption model are shown in blue line. Insets are spectrally resolved transmittance curves at 1 MV cm^{-1} in red, 0.4 MV cm^{-1} in green, 0.1 MV cm^{-1} in blue for Bi_2Se_3 films with (a) 120 QL and (b) 60 QL, and (c) $(Bi_{0.9}In_{0.1})_2Se_3$ films.

Figure 2.11 shows the integrated transmittance at room temperature. The spectra are normalized to those of substrate. The transmittance does not change much in the region between 0.1 and 50 kV cm^{-1}. From 50 kV cm^{-1} to 1 MV cm^{-1}, the transmittance increases monotonously. Above this region, it starts to saturate to 70% for 60 QL films and 63% for 120 QL films. The enhanced transparency is about 20% compared to the value in the linear region.

As shown in the insets (a) and (b) of Figure 2.11, transmittance of both films slightly decreases when the frequency reaches to zero, which can be explained by free electron Drude absorption. It is mainly related to Dirac surface states [41] and a possible slightly contribution from a two dimensional massive electron gas originated from band bending effects, especially at room temperature.[42] The absorption near 1.8 THz is related to the presence of the α-phonon.

A phenomenological saturable absorption model is used to describe the integrated transmittance versus terahertz electric field amplitude E_0: $T(E_0) = T_{ns} \frac{In[1+T_{lin}/T_{ns}(e^{\frac{E_0^2}{E_{sat}^2}}-1)]}{\frac{E_0^2}{E_{sat}^2}}$, where T_{lin} is linear integrated transmittance and T_{ns} is the non-saturable integrated transmittance, and E_{sat} is the terahertz electric field saturation value. Giorgianni *et al.* fix T_{lin}=0.59% and T_{ns} =0.69% for 60 QL films and T_{lin}=0.51% and T_{ns} =0.65% for 120 QL films and fit the data using the model above.[38] They found that E_{sat}=0.32 MV cm^{-1} for 120 QL films and E_{sat}=0.31 MV cm^{-1} for 60 QL films. These two values are comparable for both the samples. It indicates that the nonlinear absorption in topological insulators is a surface property. It is not related to bulk characteristics.

Optical response of indium doped Bi_2Se_3 films (In=10%) is investigated to assign the nonlinear electromagnetic response to Dirac electrons. Indium substitution at such level does not change the crystal structure, but it induces a phase transition. A gas of massive electrons for $(Bi_{0.9}In_{0.1})_2Se_3$ has a surface density $n_M=2.5\times10^{13}$ cm^{-2}.[38] Inset (c) of Figure 2.11 shows a flat integrated transmittance of indium doped Bi_2Se_3 films. The values of spectrally resolved transmittances at 0.1 V cm^{-1} and 1 MV cm^{-1} are almost the same. Therefore, the obvious nonlinear electromagnetic response in topological insulator Bi_2Se_3 films can be attributed to two dimensional gas of Dirac electrons present at the surface. Compared to graphene, the electromagnetic induced transparency can be interpreted by two mechanisms. One is harmonic generation, the other is a strong decrease of carrier mobility.[43]

2.5 Dirac plasmons

Plasmonics is related to light modulation via collective charge oscillations. The plasmons of ordinary massive electrons are the basic ingredients of research in optical metamaterials.[44] Plasmons of two-dimensional massless electrons in graphene exhibit an unusual behavior which enables new tunable plasmonic metamaterials as well as optoelectronic applications in the terahertz frequency range.[45] Note that plasmons cannot be excited directly by electromagnetic radiation due to dispersion law, which prevent the conservation of momentum in the photon absorption process. One way to observe the two dimensional plasmon is to pattern the surface with a subwavelength grating.[46,47]

In topological insulators, the electrons on the topological surface states are protected from backscattering, therefore the plasmon mode of topological surface states electrons shows a notably low damping character.

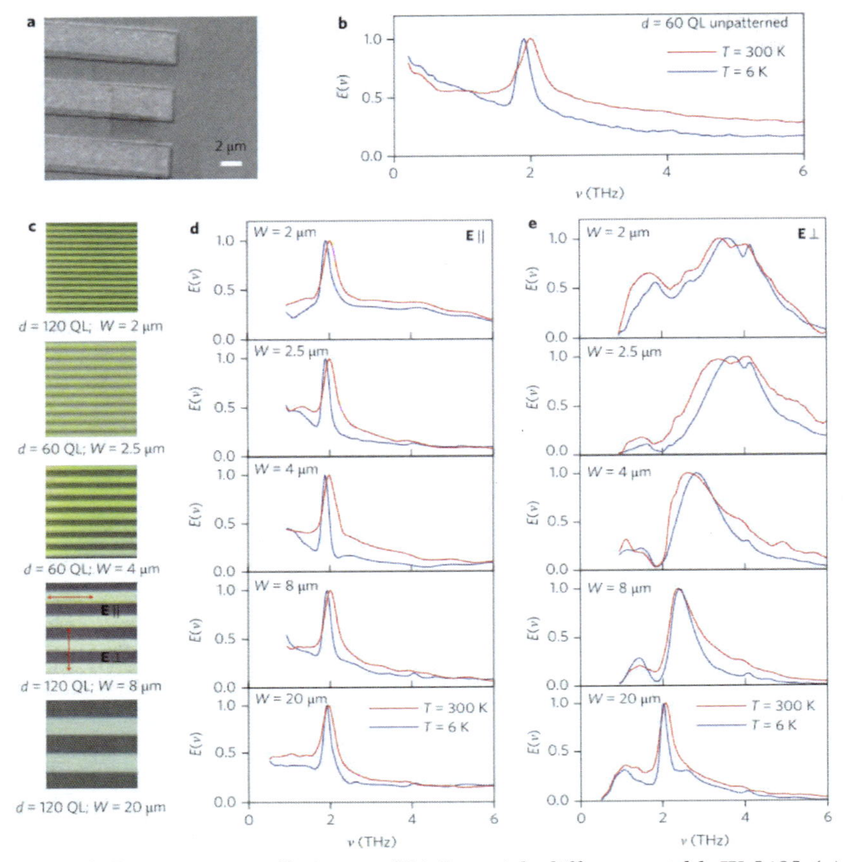

Figure 2.12 Extinction coefficients of Bi_2Se_3 with different width W.[48] (a) Scanning electron microscope image of the 2.5 μm width pattern Bi_2Se_3 films. (b) Extinction coefficient E(v) [E(v)=1-T(v)] of the as-grown unpattern films with thickness of 60 QL measured at 6 K and 300 K. It is normalized by the respective peak values. (c) Optical microscope images of pattern Bi_2Se_3 films with widths from 2 to 20 μm. Red arrows are the direction of the radiation electric field E, which is perpendicular or parallel to the ribbons. (d) Extinction coefficient of patterned films at different temperatures. The radiation electric field E is parallel and (e) perpendicularly to the ribbons.

Pietro *et al.* reported experimental evidence of plasmonic excitations in Bi_2Se_3 films on Al_2O_3 substrate using a Bruker IFS-66V Michelson interferometer equipped with a liquid helium cooled bolometer. [48] The absorption spectra were measured in the terahertz range. Bi_2Se_3 ribbons were prepared by electron beam lithography and subsequent reactive ion etching with different widths W and periods $2W$ in order to select suitable values of the plasmon wave vector $k \sim \pi/W$. They found that the linewidth of the plasmon remains almost constant in the temperature region between 6 and 300 K. Changing widths W and measuring the plasmon frequency in the terahertz range as a function of wave vector k, the dispersion curve is in good agreement with prediction for Dirac plasmons. Two peaks can be observed in Figure 2.12 (b). They are α mode at 1.85 THz and β mode at 4 THz. After patterning procedure, the spectra shown in Figure 2.12 (d) are similar to those in Figure 2.12 (b). This means the physical properties of the samples do not affect by the patterning procedure. The plasmon is observed in Figure 2.12 (e), which shows radiation electric field E is perpendicular to the ribbons. This direction is parallel to the reciprocal lattice vectors needed for energy momentum conservation.[48]

Since the peaks of the phonon modes are strongly depend on widths W, these features are related to interaction via a Fano interference between phonon and plasmon. This renormalizes the frequencies of phonon and plasmon. It hardens the mode at higher frequency and softens the mode at lower frequency. Similar phenomena is observed in doped graphene by Fei *et al.*, who reports infrared nanoscopy of two dimensional plasmon excitations of Dirac fermions, achieved by confining mid infrared radiation at the apex of a nanoscale tip.[49] The data in Figure 2.12 (e) are fitted by the a parameter-free modeling approach:[50] $E(v') = \frac{(v'+q(v'))^2}{v'^2+1} \frac{g^2}{1+(\frac{v-v_p}{\Gamma_{p/2}})^2}$. v' is renormalized frequency, which can be written as $v' = \frac{v-v_{ph}}{\frac{\Gamma_{ph(v)}}{2}} - \frac{v-v_p}{\Gamma_{p/2}}$. T Γ_{ph} is plasmon-coupled phonon linewidth: $\Gamma_{ph}(v) = \frac{2\pi v^2}{1+(\frac{v-v_p}{\Gamma_{p/2}})^2}$. q is the Fano factor, which is the ratio between the probability amplitude of exciting a discrete state (phonon) and of exciting a continuum or quasi-continuum state (plasmon), by $q(v) = \frac{\frac{vw}{g}}{\frac{\Gamma_{ph(v)}}{2}} + \frac{v-v_p}{\Gamma_{p/2}}$. Γ_p is the linewidth of plasmon, v_p is the plasmon frequency, w is the coupling factor of the radiation with the phonon and g is the coupling factor of the radiation with the plasmon.

The phonon frequency is not affected by the width W, while v_p and Γ_p decrease with the increase of width W. Γ_p is related to the Drude linewidth, the Landau damping rate, radiative decay into photons, and finite size effects. [51] Drude linewidth is independent of width, and the latter three effects are responsible for the behavior. The observation of Dirac plasmon excitation in Bi_2Se_3 by Pietro *et al.*[48] is assigned to Dirac quasiparticles of the conducting 2D edge state, based on the comparison with the theoretical dispersion relation.

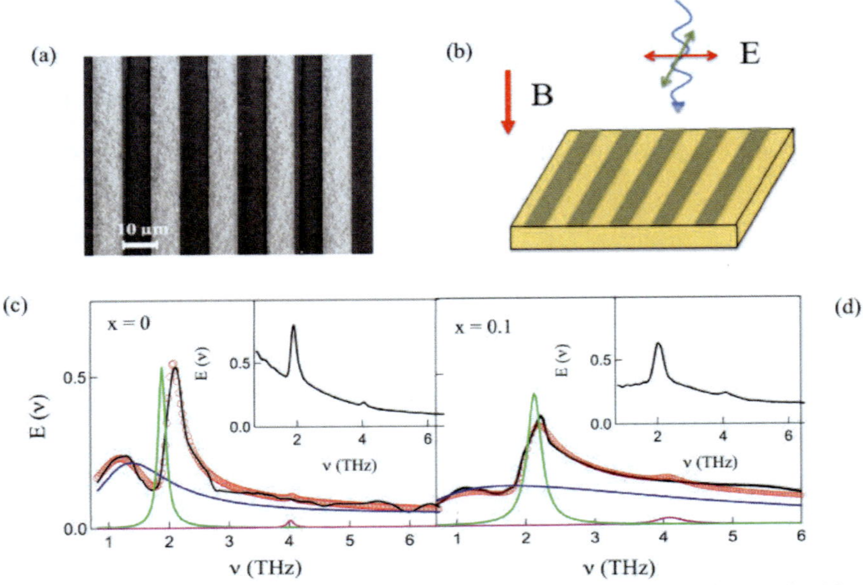

Figure 2.13 (a) Scanning electron microscopy of the patterned Bi_2Se_3 films. (b) Sketch of the magneto optic measurement configuration. Extinction coefficient spectra for patterned (c) Bi_2Se_3 and (d) $(Bi_{0.9}In_{0.1})_2Se_3$ with light polarized perpendicular to the ribbons at 1.6 K. The red empty circles are Fano fit. The solid blue lines are bare plasmon, the green lines are α phonon, and the purple lines are β phonon. The insets show the extinction spectra of unpattern samples.[53] The plasmon mode in is strongly overdamped, indicating the fundamental role of the topological protection of Dirac electrons in the plasmon lifetime. The Drude and the plasmon response of Bi_2Se_3 at low temperature are related to Dirac carriers, while it is only due to the massive surface electrons in indium doped Bi_2Se_3 with indium concentration of 10%.

The modulation of plasmons under magnetic field attracts interesting and magneto optical effects are usually related to plasmonic excitations.[52] Autore *et al.* investigated the plasmon and the Drude response as a function of external magnetic field in indium doped Bi_2Se_3.[53] They found that the cyclotron resonance can be separated from the magneto plasmon mode in energy even at low magnetic field. With increase of magnetic field, two excitations asymptotically converge at the same energy. The achievement of magnetic control of plasmonic excitations is important for plasmon controlled THz magneto optics.

2.6 Dual topological insulator

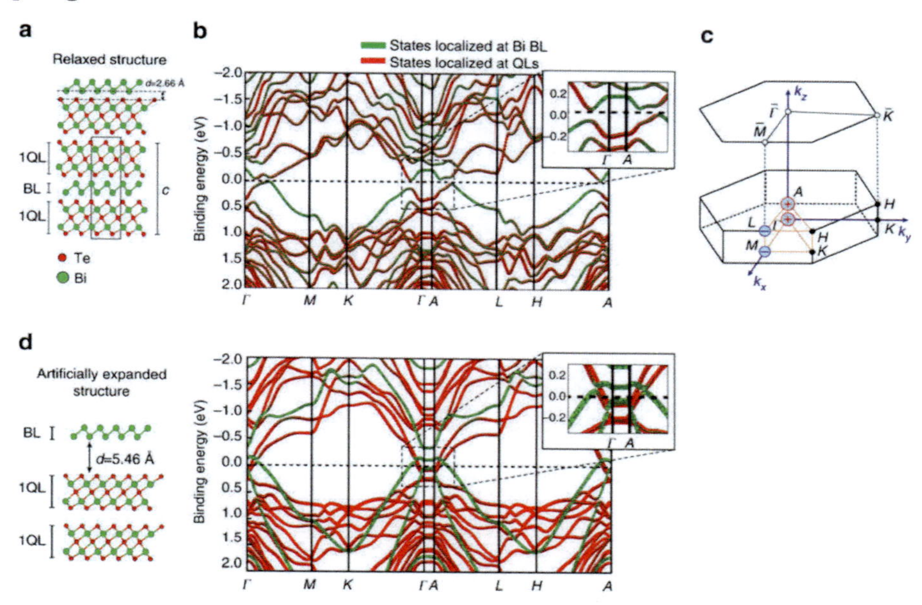

Figure 2.14 (a) Crystal structure of Bi_1Te_1. (b,d) The bulk band structure calculation in the structurally relaxed geometry (c) and with artificially expanded distances between the Bi bilayer and the QLs (d), respectively. Green lines show states localized mostly in the Bi bilayer, and the red lines show the states localized mostly in the 2QLs. Note that the band structure of the BL shows an inverted gap about 0.2 eV above the Fermi level (dashed line), as shown in (d). (c) Bulk and surface Brillouin zone with parity product of the time-reversal invariant momenta points for the relaxed structure resulting in +1 (red '+') or −1 (blue '−'). The kz direction is the stacking direction.[54]

Bi_2Te_3 was predicted to be both a strong topological insulator (STI) and a topological crystalline insulator (TCI). In a STI, time-reversal symmetry protects the metallic surface states on all surfaces. In comparison, weak topological insulator displays protected metallicity only at surfaces with a certain orientation. Other surfaces in weak topological insulator (WTI) do not contain topologically protected surface states. TCI has metallic surface states with quadratic band degeneracy on high symmetry crystal surfaces. These TCIs are the counterpart of topological insulators in materials without spin-orbit coupling. Their band structures are characterized by new topological invariants.

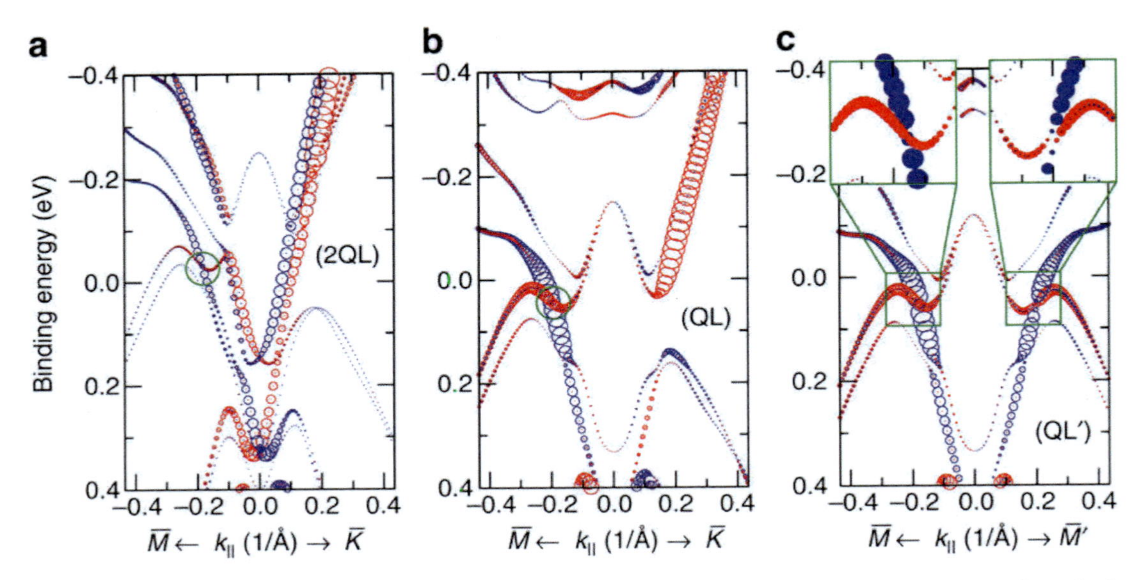

Figure 2.15 (a) Band structure of Bi_1Te_1 terminated by (a) 2QLs and (b,c) 1QL. Red symbol indicates the orientation of the spins with respect to a direction perpendicular to the momentum, with blue ones to the surface normal and green to the Dirac cone.[54]

Bi_2Te_3 is regarded as a dual TI because it exhibits two topological properties, thus it was termed a dual TI. It is simultaneously in a Z2 topological phase with Z2 invariants (v0;v1v2v3)=(1;000) and in a topological crystalline phase with mirror Chern number -1.[42] Such a combination means the possibility that controlled symmetry breaking would destroy certain surface states while keeping others intact. For instance, one could imagine a material being both a WTI and a TCI, with all its surfaces covered by metallic surface states. That is to say, the mirror plane of the TCI is normal to the dark side of the WTI. In that case, a magnetic field can destroy the topological protection of the states originated from the WTI character, while the mirror-symmetry-protected states remain intact so the dark surface is metallic.

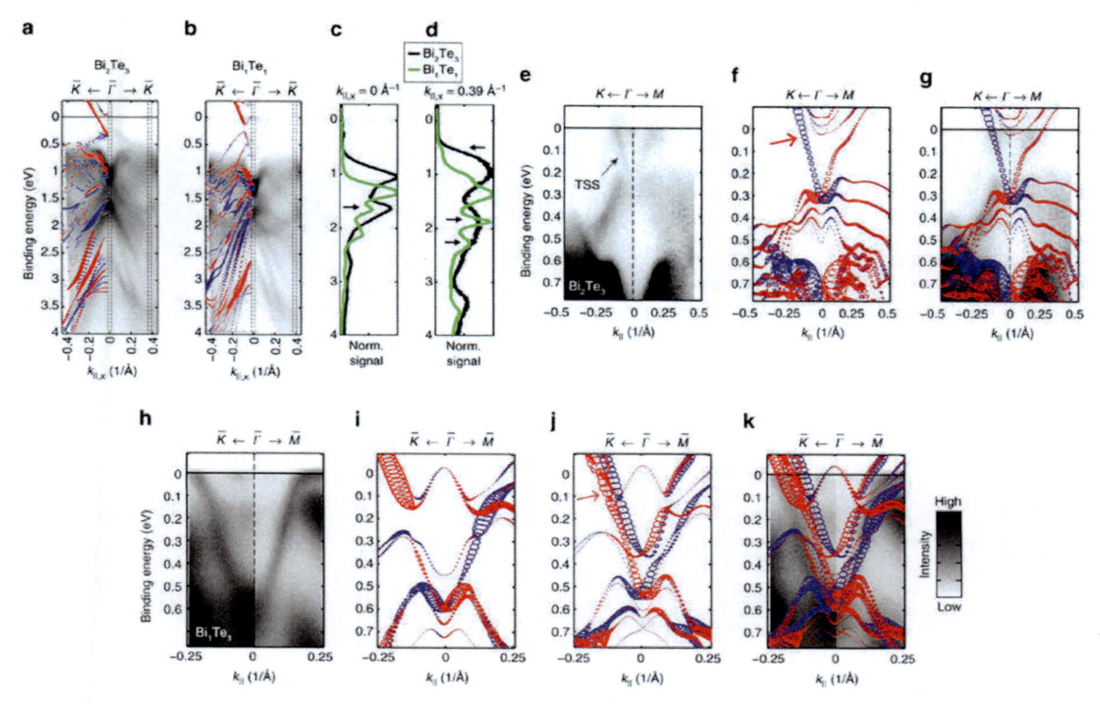

Figure 2.16 ARPES spectra of (a) Bi_2Te_3 films and (b) Bi_1Te_1 films and (c,d) their corresponding energy distribution curves. Magnified electronic structures of (e) Bi_2Te_3 films and (f) Bi_1Te_1 films. Their corresponding calculations for (f) Bi_2Te_3 films, (i) 1-QL Bi_1Te_1 films, and (j) 2-QL Bi_1Te_1 films as well as (g,k) the experimental spectra.[54]

Natural superlattices $[Bi_2]_x[Bi_2Te_3]_y$ can be fabricated in a wide range of x and y, derived from stacking hexagonal Bi_2 and Bi_2Te_3 blocks. The end members of $[Bi_2]_x[Bi_2Te_3]_y$ include metallic Bi and semiconducting Bi_2Te_3. It changes from n type for Bi_2Te_3 to p type for phases rich in Bi_2 blocks with some Bi_2Te_3, and to n type for Bi metal.[53] The unit cell of Bi_2Te_3 only consists of QL building blocks, while that of Bi_1Te_1 shows a stacking sequence of a single Bi bilayer interleaved with two subsequent QLs. Due to the different size of the unit cell along the c direction, it is easy to distinguish these compounds in the diffraction experiment.

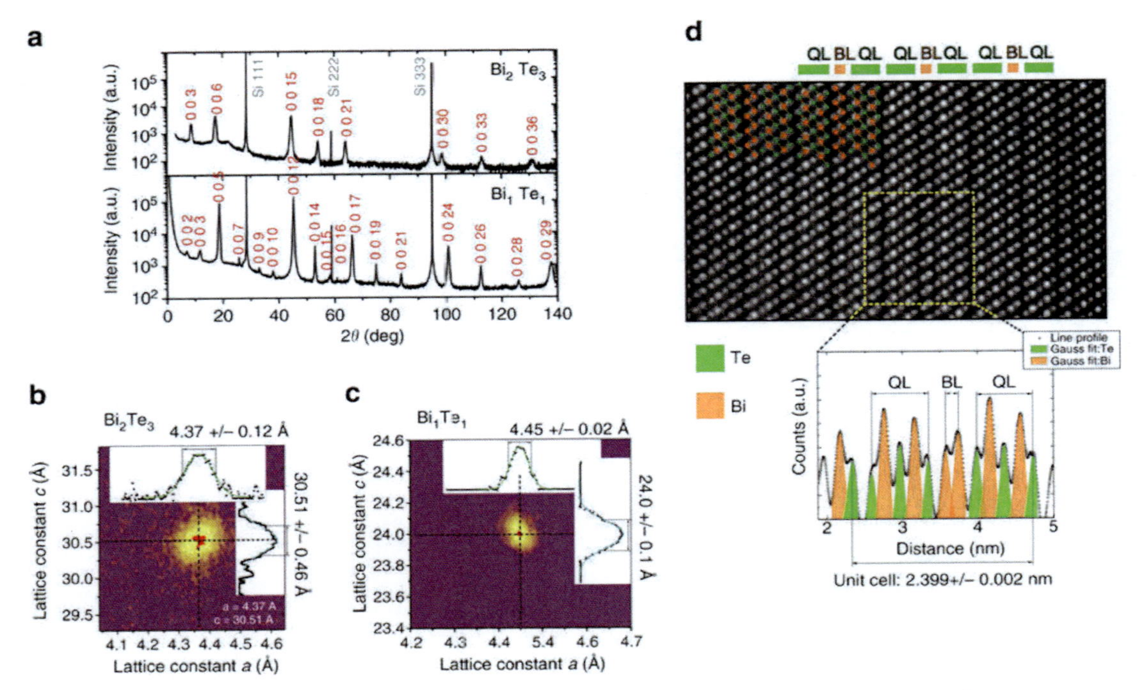

Figure 2.17 (a) XRD pattern of Bi_2Te_3 and Bi_1Te_1. Two demensional reciprocal space maps (b) around the (1,0,−1,20) reflection for Bi_2Te_3 and (c) the (1,0,−1,16) reflection for Bi_1Te_1. (d) STEM image of Bi_1Te_1.[54]

Eschbach *et al.* grow Bi_1Te_1 thin films on Si(111) by molecular beam epitaxy. They reported the stoichiometric natural superlattice Bi_1Te_1 with a small bandgap of about 0.1 eV.[54] It is a WTI and a TCI. Their density functional theory (DFT) calculations predict a Z2 class of (0;001) and a mirror Chern number -2. Two characteristic surface states of the TCI are found on the (0001) surface. Figure 2.14a shows a crystal structure of Bi_1Te_1, where Bi atom is in green and Te in orange. The separation of the layered structure into QLs and Bi bilayers is marked. Figure 2.14b shows the bulk band structure of Bi_1Te_1 in the relaxed structural geometry. Figure 2.14c shows the calculated parity of the states, indicating Bi_1Te_1 a weak TI with the (0001) surface. The calculated band structure is presented in Fig. 2.14d, introducing artificially expanded distances between the Bi bilayer and the QLs. Figure 2.15 depicts the spin-resolved surface band structure for the 1QL and 2QL surface terminations. The band structure is compatible although the surface states exist. The QL-terminated surface is near the Fermi level, protected by mirror symmetry. An upward-dispersing, Rashba-type spin-split parabolic surface band can be observed in 2 QL films. All terminations have strongly spin-polarized surface states around 1.0 eV, as shown in Figure 2.13. They are similar to the Rashba-type surface states of Bi_2Te_3 (ref. [55]) or Sb_2Te_3 (ref. [56]). The XRD pattern of

Bi_1Te_1 thin films is shown in Figure 2.16a–c and STEM in Fig. 2.16d. The peaks were fitted with Gaussians. The lattice constants a=4.37±0.12 Å and c=30.51±0.46 Å for Bi_2Te_3 and a=4.45±0.02 Å and c=24.0±0.1 Å for Bi_1Te_1. A dark field image of Bi_1Te_1 film is recorded by STEM in Figure 2.17. The contrast is associated with the difference between individual atomic columns of Bi and Te, as Bi atomic columns are brighter than Te columns. Eschbach *et al.* predicted the dual TI character of the Bi_1Te_1 by density functional theory and demonstrated by ARPES.[54] Bi_1Te_1 has a dark surface perpendicular to the stacking direction. Moreover, this dual WTI and TCI character leads to the existence of topological states on every surface of the crystal. They are protected by time reversal or by mirror symmetries.

Reference

1. A. A. Taskin, Kouji Segawa, and Yoichi Ando, Oscillatory angular dependence of the magnetoresistance in a topological insulator $Bi_{1-x}Sb_x$, Physical Review B 82, 121302(R) (2010).

2. Zhi Ren, A. A. Taskin, Satoshi Sasaki, Kouji Segawa, and Yoichi Ando, Large bulk resistivity and surface quantum oscillations in the topological insulator Bi_2Te_2Se, Physical Review B 82, 241306(R) (2010).

3. Liang Fu, Hexagonal Warping Effects in the Surface States of the Topological Insulator Bi_2Te_3, Physical Review Letters 103, 266801 (2009).

4. Tong Zhang, Peng Cheng, Xi Chen, Jin-Feng Jia, Xucun Ma, Ke He, Lili Wang, Haijun Zhang, Xi Dai, Zhong Fang, Xincheng Xie, and Qi-Kun Xue, Experimental Demonstration of Topological Surface States Protected by Time-Reversal Symmetry, Physical Review Letters 103, 266803 (2009).

5. S. Souma, K. Kosaka, T. Sato, M. Komatsu, A. Takayama, T. Takahashi, M. Kriener, Kouji Segawa, and Yoichi Ando, Direct Measurement of the Out-of-Plane Spin Texture in the Dirac-Cone Surface State of a Topological Insulator, Physical Review Letters 106, 216803 (2011).

6. Yoichi Ando, Topological insulator materials, Journal of Physical Society of Japan 82, 102001 (2013).

7. Dong-Xia Qu, Y. S. Hor, Jun Xiong, R. J. Cava, N. P. Ong, Carrier Mobility in Topological Insulators, Science 329, 821 (2010).

8. Hailin Peng, Keji Lai, Desheng Kong, Stefan Meister, Yulin Chen, Xiao-Liang Qi, Shou-Cheng Zhang, Zhi-Xun Shen, and Yi Cui, Aharonov–Bohm interference in topological insulator nanoribbons, Nature Materials 8, 225 (2010).

9. Desheng Kong, Yulin Chen, Judy J. Cha, Qianfan Zhang, James G. Analytis, Keji Lai, Zhongkai Liu, Seung Sae Hong, Kristie J. Koski, Sung-Kwan Mo, Zahid Hussain, Ian R. Fisher, Zhi-Xun Shen, and Yi Cui, Ambipolar field effect in the ternary topological insulator $(Bi_xSb_{1-x})_2Te3$ by composition tuning, Nature Nanotechnology 6, 705 (2011).

10. Zhi Ren, A. A. Taskin, Satoshi Sasaki, Kouji Segawa, and Yoichi Ando, Observations of two-dimensional quantum oscillations and ambipolar transport in the topological insulator Bi_2Se_3 achieved by Cd doping, Physical Review B 84, 075316 (2011).

11. Yong Wang, Faxian Xiu, Lina Cheng, Liang He, Murong Lang, Jianshi Tang, Xufeng Kou, Xinxin Yu, Xiaowei Jiang, Zhigang Chen, Jin Zou, and Kang L. Wang, Gate-Controlled Surface Conduction in Na-Doped Bi_2Te_3 Topological Insulator Nanoplates, Nano Letters 12 1170 (2012).

12. Yao-Yi Li, Guang Wang, Xie-Gang Zhu, Min-Hao Liu, Cun Ye, Xi Chen, Ya-Yu Wang, Ke He, Li-Li Wang, Xu-Cun Ma, Hai-Jun Zhang, Xi Dai, Zhong Fang, Xin-Cheng Xie, Ying Liu, Xiao-Liang Qi, Jin-Feng Jia, Shou-Cheng Zhang, and Qi-Kun Xue, Intrinsic topological insulator Bi_2Te_3 thin films on Si and their thickness limit, Advanced Materials 22, 4002 (2010).

13. J. Sánchez-Barriga, A. Varykhalov, G. Springholz, H. Steiner, R. Kirchschlager, G. Bauer, O. Caha, E. Schierle, E. Weschke, A. A. Ünal, S. Valencia, M. Dunst, J. Braun, H. Ebert, J. Minár, E. Golias, L. V. Yashina, A. Ney, V. Holý, and O. Rader, Nonmagnetic band gap at the Dirac point of the magnetic topological insulator, Nature Communications 7, 10559 (2016).

14. Liang He, Xufeng Kou, and Kang L. Wang, Review of 3D topological insulator thin-film growth by molecular beam epitaxy and potential applications, Physica Status Solidi Rapid Research Letters 7, 50 (2013).

15. Zhiyong Wang, Tao Lin, Peng Wei, Xinfei Liu, Randy Dumas, Kai Liu, and Jing Shi, Tuning carrier type and density in Bi_2Se_3 by Ca-doping, Applied Physics Letters 97, 042112 (2010).

16. J. Navrátil, J. Horák, T. Plechacek, S. Kamba, P. Lošt'ák, J. S. Dyck, W. Chen, and C. Uher, Conduction band splitting and transport properties of Bi_2Se_3, Journal of Solid State Chemistry 177, 1704 (2004).

17. P. A. Sharma, A. L. Lima Sharma, M. Hekmaty, K. Hattar, V. Stavila, R. Goeke, K. Erickson, D. L. Medlin, M. Brahlek, N. Koirala, and S. Oh, Ion beam modification of topological insulator bismuth selenide, Applied Physics Letters 105, 242106 (2014).

18. Y. S. Hor, A. Richardella, P. Roushan, Y. Xia, J. G. Checkelsky, A. Yazdani, M. Z. Hasan, N. P. Ong, and R. J. Cava, p-type Bi_2Se_3 for topological insulator and low-temperature thermoelectric applications, Physical Review B 79, 195208 (2009).

19. Panju Shang, Xin Guo, Bao Zhao, Xianqi Dai, Li Bin, Jinfeng Jia, Quan Li, and Maohai Xie, Nanoclusters of CaSe in calcium-doped Bi_2Se_3 grown by molecular-beam epitaxy, Nanotechnology 27, 085601 (2016).

20. Y. L. Chen, J.-H. Chu, J. G. Analytis, Z. K. Liu, K. Igarashi, H.-H. Kuo, X. L. Qi, S. K. Mo, R. G. Moore, D. H. Lu, M. Hashimoto, T. Sasagawa, S. C. Zhang, I. R. Fisher, Z. Hussain, and Z. X. Shen, Massive Dirac fermion on the surface of a magnetically doped topological insulator, Science 329, 659 (2010).

21. Masatoshi Sato and Yoichi Ando, Topological Superconductors, arXiv:1608.03395.

22. Michael Schecter, Karsten Flensberg, Morten H. Christensen, Brian M. Andersen, and Jens Paaske, Self-organized topological superconductivity in a Yu-Shiba-Rusinov chain, Physical Review B 93, 140503(R) (2016).

23. Masatoshi Sato and Satoshi Fujimoto, Majorana Fermions and Topology in Superconductors, Journal of the Physical Society of Japan 85, 072001 (2016).

24. Pablo San-Jose, Jorge Cayao, Elsa Prada, and Ramón Aguado, Majorana bound states from exceptional points in non-topological superconductors, Scientific Reports 6, 21427 (2016).

25. V. Kaladzhyan, C. Bena, and P. Simon, Asymptotic behavior of impurity-induced bound states in low-dimensional topological superconductors, Journal of Physics: Condensed Matter 28 485701 (2016).

26. R. S. Akzyanov, A. L. Rakhmanov, A. V. Rozhkov, and Franco Nori, Tunable Majorana fermion from Landau quantization in 2D topological superconductors, Physical Review B 94, 125428 (2016).

27. Stefano Valentini, Michele Governale, Rosario Fazio, and Fabio Taddei, Physica E: Low-dimensional Systems and Nanostructures 75, 15 (2016).

28. S. A. Jafari and Farhad Shahbazi, Exactly solvable spin chain models corresponding to BDI class of topological superconductors, Scientific Reports 6, 32720 (2016).

29. Yukio Tanaka, Masatoshi Sato, and Naoto Nagaosa, Symmetry and Topology in Superconductors –Odd-Frequency Pairing and Edge States, Journal of the Physical Society of Japan 81 011013 (2012).

30. Y. S. Hor, A. J. Williams, J. G. Checkelsky, P. Roushan, J. Seo, Q. Xu, H. W. Zandbergen, A. Yazdani, N. P. Ong, and R. J. Cava, Superconductivity in $Cu_xBi_2Se_3$ and its Implications for Pairing in the Undoped Topological Insulator, Physical Review Letters 104, 057001 (2010).

31. Matthew Brahlek, Namrata Bansal, Nikesh Koirala, Su-Yang Xu, Madhab Neupane, Chang Liu, M. Zahid Hasan, and Seongshik Oh, Topological-metal to band-insulator transition in $(Bi_{1-x}In_x)_2Se_3$ thin films, Physical Review Letters 109, 186403 (2012).

32. Jiajun Zhu, Fang Liu, Shengqiang Zhou, C. Franke, S. Wimmer, V. V. Volobuev, G. Springholz, A. Pashkin, H. Schneider, and M. Helm, Lattice vibrations and electrical transport in $(Bi_{1-x}In_x)_2Se_3$ films, Applied Physics Letters 109, 202103 (2016).

33. Rui Lou, Zhonghao Liu, Wencan Jin, Haifeng Wang, Zhiqing Han, Kai Liu, Xueyun Wang, Tian Qian, Yevhen Kushnirenko, Sang-Wook Cheong, Richard M. Osgood, Jr., Hong Ding, and Shancai Wang, Sudden gap closure across the topological phase transition in $Bi_{2-x}In_xSe_3$, Physical Review B 92, 115150 (2015).

34. M. I. Daunov, I. K. Kamilov, and S. F. Gabibov, Concept of mobility threshold: The Ioffe-Regel rule, Physics of the Solid State 52, 2019 (2010).

35. Maryam Salehi, Hassan Shapourian, Nikesh Koirala, Matthew J. Brahlek, Jisoo Moon, and Seongshik Oh, Finite-size and composition-driven topological phase transition in $(Bi_{1-x}In_x)_2Se$ thin films, Nano Letters 16, 5528 (2016).

36. Marta Autore, Flavio Giorgianni, Fausto D' Apuzzo, Alessandra Di Gaspare, Irene Lo Vecchio, Matthew Brahlek, Nikesh Koirala, Seongshik Oh, Urlich Schade, Michele Ortolanic, and Stefano Lupi, Topologically protected Dirac plasmons and their evolution across the quantum phase transition in a $(Bi_{1-x}In_x)_2Se$ topological insulator, Nanoscale 8, 4667 (2016).

37. Liang Wu, M. Brahlek, R. Valdés Aguilar, A. V. Stier, C. M. Morris, Y. Lubashevsky, L. S. Bilbro, N. Bansal, S. Oh and N. P. Armitage, A sudden collapse in the transport lifetime across the topological phase transition in $(Bi_{1-x}In_x)_2Se$, Nature Physics 9, 410 (2013).

38. Flavio Giorgianni, Enrica Chiadroni, Andrea Rovere, Mariangela Cestelli-Guidi, Andrea Perucchi, Marco Bellaveglia, Michele Castellano, Domenico Di Giovenale, Matthew Brahlek, Nikesh Koirala, Seongshik Oh, and Stefano Lupi, Strong nonlinear terahertz response induced by Dirac surface states in Bi_2Se_3 topological insulator, Nature Communications 7, 11421 (2016).

39. E. Chiadroni, M. Bellaveglia, P. Calvani, M. Castellano, L. Catani, A. Cianchi, G. Di Pirro, M. Ferrario, G. Gatti, O. Limaj, S. Lupi, B. Marchetti, A. Mostacci, E. Pace, L. Palumbo, C. Ronsivalle, R. Pompili, and C. Vaccarezza, Characterization of the THz radiation source at the Frascati linear accelerator, Review of Scientific Instruments 84, 022703 (2013).

40. E. Chiadroni, A. Bacci, M. Bellaveglia, M. Boscolo, M. Castellano, L. Cultrera, G. Di Pirro, M. Ferrario, L. Ficcadenti, D. Filippetto, G. Gatti, E. Pace, A. R. Rossi, C. Vaccarezza, L. Catani, A. Cianchi, B. Marchetti, A. Mostacci, L. Palumbo, C. Ronsivalle, A. Di Gaspare, M. Ortolani, A. Perucchi, P. Calvani, O. Limaj, D. Nicoletti, and S. Lupi, The SPARC linear accelerator based terahertz source, Applied Physics Letters 102, 09410 (2013).

41. R. Valdés Aguilar, A. V. Stier, W. Liu, L. S. Bilbro, D. K. George, N. Bansal, L. Wu, J. Cerne, A. G. Markelz, S. Oh, and N. P. Armitage, Terahertz response and colossal kerr rotation from the surface states of the topological insulator Bi_2Se_3, Physical Review Letters 108, 087403 (2012).

42. T. Stauber, G. Gómez-Santos, and L. Brey, Spin-charge separation of plasmonic excitations in thin topological insulators, Physical Review B 88, 205427 (2013).

43. P. Bowlan, E. Martinez-Moreno, K. Reimann, T. Elsaesser, and M. Woerner, Ultrafast terahertz response of multilayer graphene in the nonperturbative regime, Physical Review B 89, 041408 (2014).

44. S. A. Maier, Plasmonics: fundamentals and applications, Springer (2007).

45. M. Jablan, H. Buljan, M. Soljacic, Plasmonics in graphene at infrared frequencies, Physical Review B 80, 245435 (2009).

46. Long Ju, Baisong Geng, Jason Horng, Caglar Girit, Michael Martin, Zhao Hao, Hans A. Bechtel, Xiaogan Liang, Alex Zettl, Y. Ron Shen, and Feng Wang, Graphene plasmonics for tunable terahertz metamaterials, Nature Nanotechnology 6, 630 (2011).

47. S. J. Allen, Jr., D. C. Tsui, and R. A. Logan, Observation of the Two-Dimensional Plasmon in Silicon Inversion Layers, Physical Review Letters 38, 980 (1977).

48. P. Di Pietro, M. Ortolani, O. Limaj, A. Di Gaspare, V. Giliberti, F. Giorgianni, M. Brahlek, N. Bansal, N. Koirala, S. Oh, P. Calvani, and S. Lupi, Observation of Dirac plasmons in a topological insulator, Nature Nanotechnology 8, 556 (2013).

49. Zhe Fei, Gregory O. Andreev, Wenzhong Bao, Lingfeng M. Zhang, Alexander S. McLeod, Chen Wang, Margaret K. Stewart, Zeng Zhao, Gerardo Dominguez, Mark Thiemens, Michael M. Fogler, Michael J. Tauber, Antonio H. Castro-Neto, Chun Ning Lau, Fritz Keilmann, and Dimitri N. Basov, Infrared Nanoscopy of Dirac Plasmons at the Graphene-SiO_2 Interface, Nano Letters 11, 4701 (2011).

50. Vincenzo Giannini, Yan Francescato, Hemmel Amrania, Chris C. Phillips, and Stefan A. Maier, Fano resonances in nanoscale plasmonic systems: a parameter-free modeling approach, Nano Letters 11, 2835 (2011).

51. Hugen Yan, Tony Low, Wenjuan Zhu, Yanqing Wu, Marcus Freitag, Xuesong Li, Francisco Guinea, Phaedon Avouris, and Fengnian Xia, Damping pathways of mid-infrared plasmons in graphene nanostructures, Nature Photonics 7, 394 (2013).

52. Gaspar Armelles, Alfonso Cebollada, Antonio García-Martín, and María Ujué González, Magnetoplasmonics: Combining Magnetic and Plasmonic Functionalities, Advanced Optical Materials 1, 10 (2013).

53. Marta Autore, Hans Engelkamp, Fausto D'Apuzzo, Alessandra Di Gaspare, Paola Di Pietro, Irene Lo Vecchio, Matthew Brahlek, Nikesh Koirala, Seongshik Oh, and Stefano Lupi, Observation of magnetoplasmons in Bi_2Se_3 topological insulator, ACS Photonics 2, 1231 (2015).

54. Markus Eschbach, Martin Lanius, Chengwang Niu, Ewa Młyńczak, Pika Gospodarič, Jens Kellner, Peter Schüffelgen, Mathias Gehlmann, Sven Döring, Elmar Neumann, Martina Luysberg, Gregor Mussler, Lukasz Plucinski, Markus Morgenstern, Detlev Grützmacher, Gustav Bihlmayer, Stefan Blügel, and Claus M. Schneider, Bi_1Te_1 is a dual topological insulator, Nature Communications 8, 14976 (2017).

55. A. Herdt, L. Plucinski, G. Bihlmayer, G. Mussler, S. Döring, J. Krumrain, D. Grützmacher, S. Blügel, and C. M. Schneider, Spin-polarization limit in Bi_2Te_3 Dirac cone studied by angle- and spin-resolved photoemission experiments and ab initio calculations, Physical Review B 87, 035127 (2013).

56. C. Pauly, G. Bihlmayer, M. Liebmann, M. Grob, A. Georgi, D. Subramaniam, M. R. Scholz, J. Sánchez-Barriga, A. Varykhalov, S. Blügel, O. Rader, and M. Morgenstern Probing two topological surface bands of Sb2Te3 by spin-polarized photoemission spectroscopy, Physical Review B 86, 235106 (2012).

Chapter 3 Raman spectroscopy

3.1 Stokes and anti-stokes scattering

Raman scattering spectroscopy is a prospective tool to study phonon modes. Since Raman spectroscopy can successfully identify the graphene layers, which has a similar band structure of topological surface states, it is a strong motivation to investigate phonon in topological insulators by Raman spectroscopy.[1-5] The primitive unit cell of Bi_2Se_3 has five atoms. Correspondingly, there are fifteen modes at the center of the Brillouin zone. Three are acoustic modes and twelve are optical modes. In these twelve optical mods, Raman-active modes are $2A_{1g}+2E_g$, and infrared-active modes are $2A_{1u}+2E_u$. [6] Figure 3.1 shows the atomic displacements of the Raman-active modes.

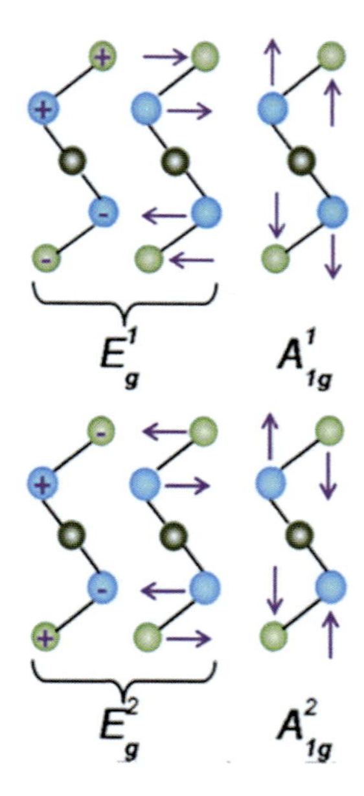

Figure 3.1 The Raman-acitve modes of Bi_2Se_3.[6]

The Raman tensors can be written as:

A_{1g}:

$$\begin{pmatrix} a & \square & \square \\ \square & a & \square \\ \square & \square & b \end{pmatrix}$$

E_g:

$$\begin{pmatrix} c & \square & \square \\ \square & -c & d \\ \square & d & \square \end{pmatrix} \quad \begin{pmatrix} \square & -c & -d \\ -c & \square & \square \\ -d & \square & \square \end{pmatrix}$$

The E_g mode is two-fold in-plane vibrational mode while the A_{1g} mode vibrates along the [0001] direction and it is out-of-plane mode. They can be distinguished by the nonzero off-diagonal components in the E_g Raman tensor.

Stokes Raman scattering means that atom or molecule absorbs energy, while atom or molecule loses energy is anti-Stokes scattering. That is to say, in Stokes Raman scattering, scattered photon has less energy than the incident photon, while in anti-Stokes Raman scattering, scattered photon has more energy than the incident photon. Figure 3.2 shows Raman spectra including both Stokes and anti-Stokes contributions in parallel polarization configuration (upper curve) and perpendicular polarization configuration (lower curve). Parallel polarization configuration corresponds to the x-x component of the Raman tensors, while perpendicular one is the x-y component of the Raman tensors.

According to the selection rule and the Raman tensors, the two peaks in the x-y polarization configuration can be assigned as E_g modes due to its containing off-diagonal elements. The other two peaks in the x-x polarization configuration is A_{1g} modes. The anti-Stokes spectra supports the assignment as well. Four peak can be clearly observed at 37, 72, 131, 174 cm^{-1}, which can be assigned as E_g^1, A_{1g}^1, E_g^2, and A_{1g}^2, respectively. These four phonon modes were firstly observed by Zhang et al. in Bi_2Se_3 nanoplatelets.[7]

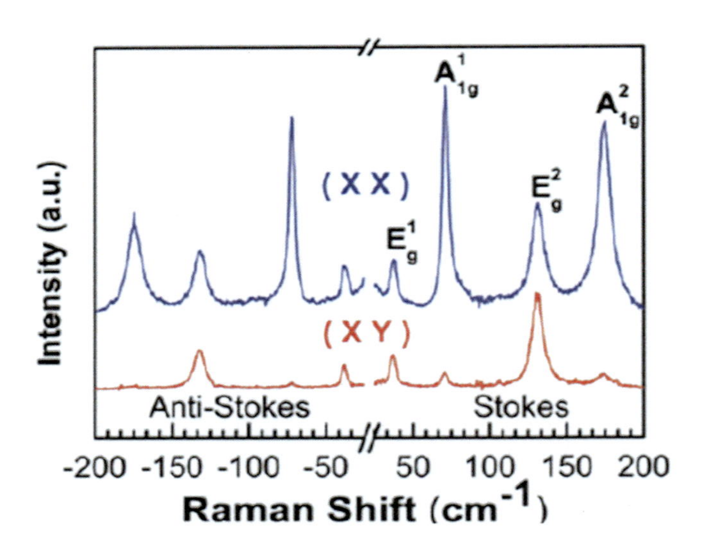

Figure 3.2 Raman spectra of Bi_2Se_3 in parallel (XX) and perpendicular (XY) polarizations. Both Stokes and anti-Stokes contributions are shown.[7]

3.2 Thickness dependence

3.2.1 Bi_2Se_3 films on GaAs

Figure 3.3 shows Raman spectra of Bi_2Se_3 films on GaAs with thickness from 1 quintuple layers to 18 ones, carried out at 80 K using a 514 nm laser excitation in the parallel polarization configuration.[8] Note that the spectra are normalized to the peak intensity of A_{1g} mode around 74 cm^{-1}. Raman spectra of 18 QLs are very similar to the bulk spectra.

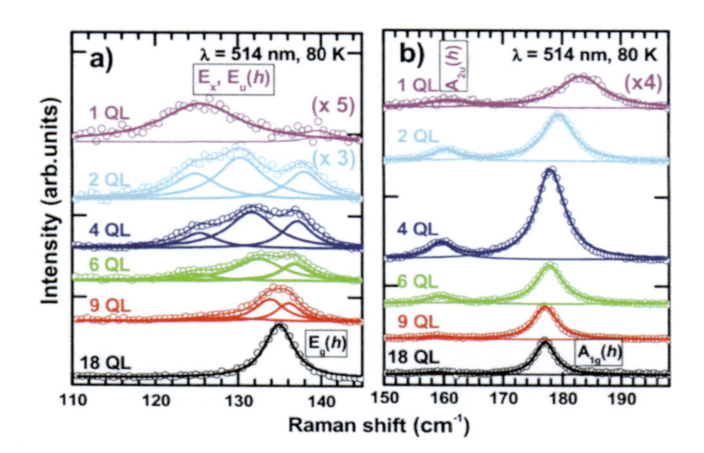

Figure 3.3 Evolution of the Raman spectra of Bi_2Se_3 films on GaAs with number of quintuple layers. Open circles are the experimental data and solid lines are the Lorentzian fitting curves.[8]

E_g mode of 18 QL can be well resolved by a single peak, which broadens with the decrease of the film thickness. It transforms into two peaks for the 9 QL films, and into three ones for the 6, 4, and 2 QL samples. Variation of film thickness changes the frequency, width, and intensity of peaks. The appearance of the other peak for 9 to 2 QL films around 130 cm^{-1} can be assigned to infrared active mode E_u.[9] And the third peak is likely due to splitting of the degeneracy of E_g or E_u modes. The frequency of the mode increases with the increase of film thickness, which is in agreement with some theoretical calculations, [10] while other calculations are totally different.[11] The additional peak at ~160 cm^{-1} appearing for thickness below 9 QL can be assigned to infrared active mode A_{2u} mode. The reduced thickness possibly induces a finite Raman active mode because of the breaking of the crystal symmetry along the growth direction.

3.2.2 Bi_2Se_3 nanoplates on graphene

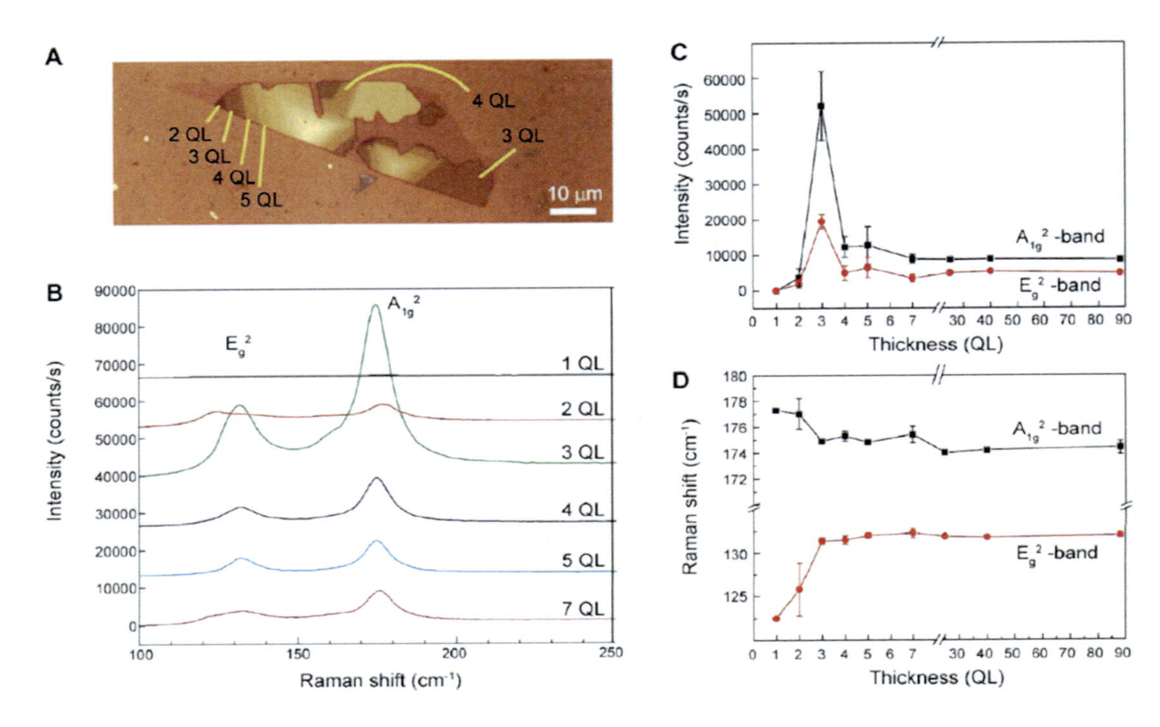

Figure 3.4 Layered Bi2Se3 nanoplates grown on a few-layer graphene substrate. Raman spectra were measured with 699 nm laser. [12]

Graphene and topological insulators are Dirac materials, attracting extensive attention due to their distinctive band structures and physical properties. In particular, nanostructure Dirac materials own extremely large surface to volume ratios and distinct edge effects, with great advantage to suppress the residual conductance in the bulk comparing to bulk materials. Interlayer interactions of Dirac fermions in Bi_2Se_3 and graphene heterostructure are an exciting new direction in theoretical and experimental research. Using a simple vapor-phase deposition method, Dang et al. reported different number (1-10 QL) of Bi_2Se_3 layers with defined orientations epitaxial grown on a few-layer graphene substrate. [12] Three quintuple layers (3-QL) sample exhibits the strongest Raman intensity while 1-QL and 2-QL samples experience tensile stress, consistent with compressive stress in monolayer and bilayer graphene.

3.3 Temperature dependence of Bi_2Se_3 crystal

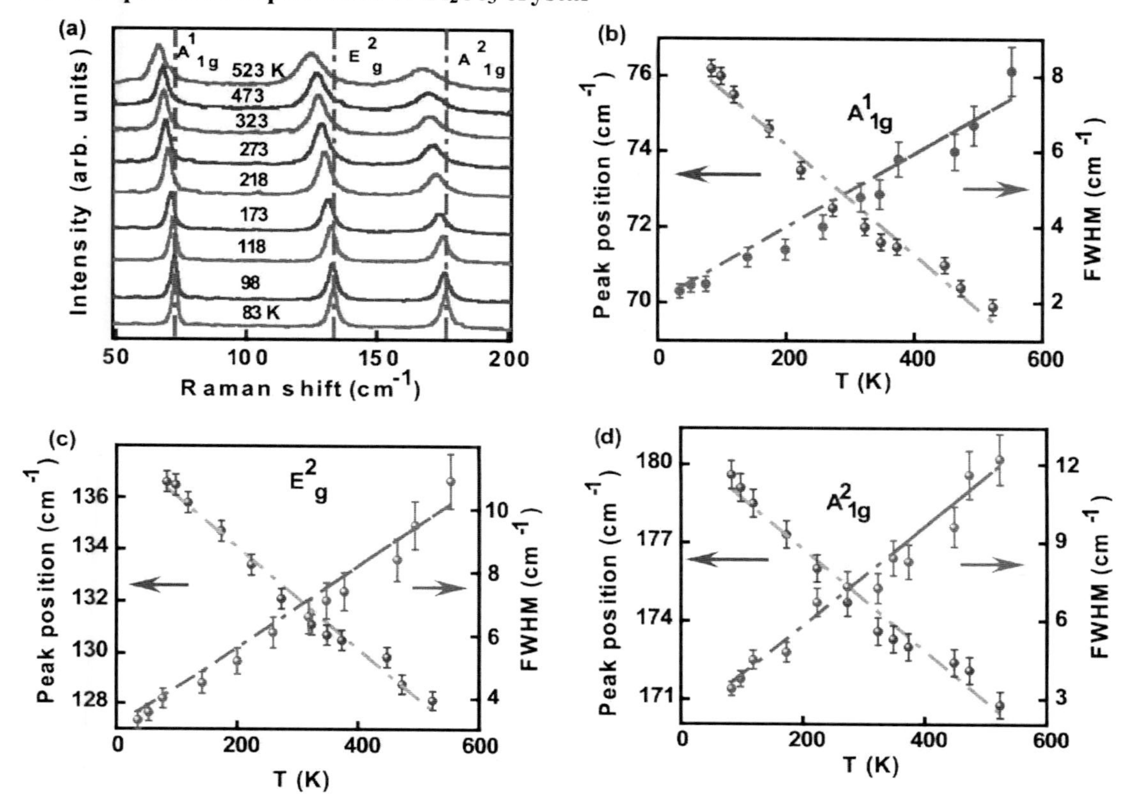

Figure 3.5 (a) Temperature dependent Raman spectra of Bi_2Se_3 crystal. (b)-(d) Peak positions and full width at half maximum (FWHM) of different Raman modes.[13]

Temperature dependent Raman spectroscopy technique is very useful to understand the lattice anharmonicity, phase transition, and spin-phonon coupling in materials. The knowledge of temperature coefficients of Raman modes can be used to estimate the thermal conductivity of materials. Irfan et al. reported a detailed and systematic temperature dependent Raman study on Bi_2Se_3 crystal from 83 to 523 K.[13] The Bi_2Se_3 crystal was grown using a modified Bridgman technique. Raman measurements were carried out using a Horiba-Yobin T64000 micro-Raman system. 532 nm wavelength laser was used as an excitation source. Figure 3.5(a) shows the temperature dependent Raman spectra of Bi_2Se_3 crystal. With increase of the temperature, the peak positions of all the phonon modes show a red-shift. Moreover, the full width at half maximum (FWHM) of all the phonon modes increases monotonically with increase of the temperature, indicating shorter phonon lifetime or phonon relaxation time τ at high temperatures. Several phonon scattering mechanisms exist. The dominant mechanisms are Umklapp process, boundary scattering, phonon-impurity scattering, and electron-phonon scattering, each of which is related to relaxation time τ, inversely proportional to the relaxation rate. The total relaxation rate of phonon is given by Matthiessens rule: $1/\tau=1/\tau_u+1/\tau_b+1/\tau_i+1/\tau_{e-ph}$. That is to say, the relaxation time τ of the optical phonon is associated with the rate of approaching equilibrium.

Raman mode	Peak position (cm^{-1})	FWHM (cm^{-1})	χ (10^{-2} cm^{-1}K^{-1})
A_{1g}^1	72	4.5	-1.44
E_{2g}^2	131	6	-1.94
A_{1g}^2	175	6.5	-1.95

Table 3.1 Raman peak position, FWHM, temperature coefficient of different phonon modes in Bi_2Se_3 crystal.[13]

A linear equation is used to fit the data $\omega(T)=\omega_o+\chi T$, where ω_o is the frequency of vibration of the phonon modes at absolute zero temperature, χ is the first order temperature coefficient of the phonon modes. The fitting parameter values are listed in Table 3.1. The value of the first order temperature coefficient is from -1.44 to -1.95 $\times 10^{-2}$ cm^{-1}K^{-1} in Bi_2Se_3 crystal. The variation in peak position of Raman mode is mainly due to the volume contribution or thermal expansion that leads to anharmonicity.

3.4 Pressure dependence of Bi_2Se_3 crystal

High pressure studies on material are extremely important to understand its properties as the increase in pressure reduces the interatomic distance and to finely tune the materials properties. Hamlin et. al. reported x-ray diffraction, electrical resistivity, and magnetoresistance measurements on Bi_2Se_3 crystal under high pressure and found that pressure induces great changes of the electrical resistivity.[14] At the beginning, pressure drives Bi_2Se_3 crystal toward increasingly insulating behavior, while at higher pressure, it appears to enter a fully metallic state coincident with a change in the crystal structure. The low pressure phase of Bi_2Se_3 shows an unusual field dependence of the transverse magnetoresistance $\Delta\rho_{xx}$ that is positive at low fields. It becomes negative at higher fields. Their results demonstrate that pressure below 8 GPa provides a non-chemical means to reduce the bulk conductivity of Bi_2Se_3. Deshpande et al. reported pressure dependent Raman spectra of Bi_2Se_3 crystal, finding a first order phase transition above 10 GPa.[15]

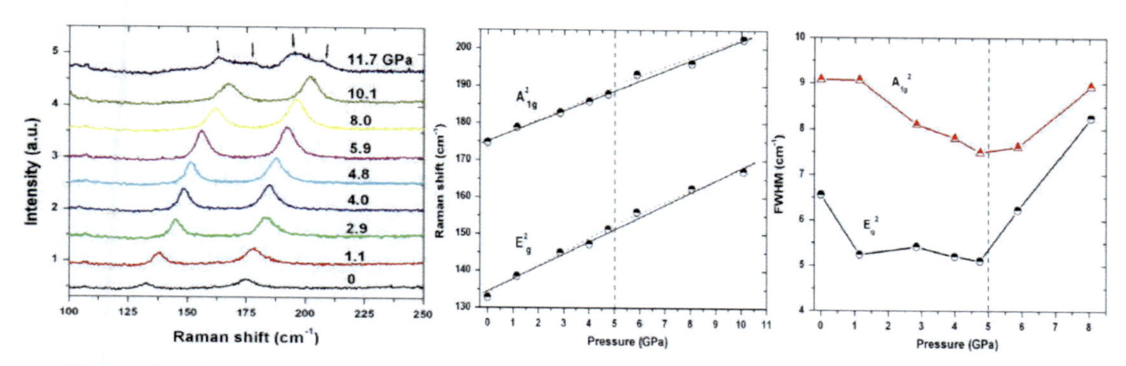

Figure 3.6 Raman spectra of Bi_2Se_3 crystal at different pressures up to 11.7 GPa. Peak position and FWHM of A_{1g}^2 and E_g^2 vs pressure.[15]

Above 10 GPa, the Raman spectra shows obvious changes, as shown in Figure 3.6. A_{1g}^2 and E_g^2 modes appear as a single broad band up to 10.1 GPa. Above it, new Raman modes starts to appear. This is due to the phase transition above 10 GPa from hexagonal to a new phase. Both Raman modes increase monotonically with increasing pressure before the phase transition. The increase obeys a linear fit, $\omega_i(P)=\omega_0+AP$. Where $A=d\omega_i/dP$ is a pressure coefficient. The fitting parameters are listed in Table 3.2. The parameters A are 3.4 and 2.69 cm^{-1}/GPa for E_g^2 and A_{1g}^2 modes, very close to theoretical results at 3.13 and 2.68 cm^{-1}/GPa.[16]

Raman mode	Peak position ω_i (cm^{-1})	A (cm^{-1}/GPa)
E_{2g}^2	132.5	3.4
A_{1g}^2	174	2.69

Table 3.2 Raman peak position and fitting parameter of different phonon modes in Bi_2Se_3 crystal at different pressures.[15]

E_g^2 and A_{1g}^2 modes are mainly associated with intralayer vibrations, therefore, their similar pressure coefficient indicates a similar covalent nature of the intralayer bonds. Pressure dependent Raman shift and linewidth show an electronic topological transition by a small change of the pressure coefficients and reduction in linewidths of both Raman mode frequencies around 5 GPa, as shown by dotted line in Figure 3.6. Manjon et al. [16] pointed out that the presence of a pressure-induced electronic topological transition in Bi_2Se_3 at low pressure is driven by the change of the Van der Waals forces. It is a second-order isostructural phase transition. Raman studies on other topological insulators, such as Bi_2Te_3 and BiTeI can be found elsewhere.[17-25]

Reference

1. V. Meunier, A. G. Souza Filho, E. B. Barros, and M. S. Dresselhaus, Physical properties of low-dimensional sp^2-based carbon nanostructures, Reviews of Modern Physics 88, 025005 (2016).

2. L. M. Malard, M. A. Pimenta, G. Dresselhaus, and M. S. Dresselhaus, Raman spectroscopy in graphene, Physics Reports 473, 51 (2009).

3. A. C. Ferrari, J. C. Meyer, V. Scardaci, C. Casiraghi, M. Lazzeri, F. Mauri, S. Piscanec, D. Jiang, K. S. Novoselov, S. Roth, and A. K. Geim, Raman Spectrum of Graphene and Graphene Layers, Physical Review Letters 97, 187401 (2006).

4. Yuyoung Shin, Marcelo Lozada-Hidalgo, Jose L. Sambricio, Irina V. Grigorieva, Andre K. Geim, and Cinzia Casiraghi, Raman spectroscopy of highly pressurized graphene membranes, Applied Physics Letters 108, 221907 (2016).

5. Xin Zhang, Weng-Peng Han, Xiao-Fen Qiao, Qing-Hai Tan, Yu-Fang Wang, Jun Zhang, and Ping-Heng Tan, Raman characterization of AB- and ABC-stacked few-layer graphene by interlayer shear modes, Carbon 99, 118 (2016).

6. Yuan Yan, Xu Zhou, Han Jin, Cai-Zhen Li, Xiaoxing Ke, Gustaaf Van Tendeloo, Kaihui Liu, Dapeng Yu, Martin Dressel, and Zhi-Min Liao, Surface-Facet-Dependent Phonon Deformation Potential in IndividualStrained Topological Insulator Bi_2Se_3 Nanoribbons, ACS Nano 9, 10244 (2015).

7. Jun Zhang, Zeping Peng, Ajay Soni, Yanyuan Zhao, Yi Xiong, Bo Peng, Jianbo Wang, Mildred S. Dresselhaus, and Qihua Xiong, Raman spectroscopy of few-quintuple layer topological insulator Bi_2Se_3 nanoplatelets, Nano Letters 11, 2497 (2011).

8. Mahmoud Eddrief, Paola Atkinson, Victor Etgens, and Bernard Jusserand, Low-temperature Raman fingerprints for few-quintuple layer topological insulator Bi_2Se_3 films epitaxied on GaAs, Nanotechnology 25, 245701 (2014).

9. W. Richter, H. Köhler, and C. R. Becker, A Raman and farinfrared investigation of phonons in the rhombohedral V2-VI3 compounds, Physca Status Solidi B 84, 619 (1977).

10. V. Chis, S. I. Yu, K. A. Kokh, V. A. Volodin, O. E. Tereshchenko, and E. V. Chulkov, Vibrations in binary and ternary topological insulators: first-principles calculations and Raman spectroscopy measurements, Physical Review B 86, 174304 (2012).

11. W. Cheng and S. F. Ren, Phonons of single quintuple Bi_2Te_3 and Bi_2Se_3 films and bulk materials, Physical Review B 83, 094301 (2011).

12. Wenhui Dang, Hailin Peng, Hui Li, Pu Wang, and Zhongfan Liu, Nano Letters 10, 2870 (2010).

13. Bushra Irfan, Satyaprakash Sahoo, Anand P. S. Gaur, Majid Ahmadi, Maxime J.-F. Guinel, Ram S. Katiyar, and Ratnamala Chatterjee, Temperature dependent Raman scattering studies of three dimensional topological insulators Bi_2Se_3, Journal of Applied Physics 115, 173506 (2014).

14. J. J. Hamlin, J. R. Jeffries, N. P. Butch, P. Syers, D. A. Zocco, S. T. Weir, Y. K. Vohra, J. Paglione and M. B. Maple, High pressure transport properties of the topological insulator Bi_2Se_3, Journal of Physics: Condensed Matter 24, 035602 (2012).

15. M. P. Deshpandea, Sandip V. Bhatta, Vasant Satheb, Rekha Raoc, and S. H. Chakia, Physica B: Condensed Matter 433, 72 (2014).

16. R. Vilaplana, D. Santamaria-Perez, O. Gomis, F.J. manjon, J. Gonzalez, A. Segura, A. Munoz, P. Rodriguez-Hernandez, E. Perez-Gonzalez, V. Marin-Borras, V. Munoz-Sanjose, C. Drasar, V. Kucek, Structural and vibrational study of Bi2Se3 under high pressure, Physical Review B, 84 184110 (2011).

17. Rui He, Zhenhua Wang, Richard L. J. Qiu, Conor Delaney, Ben Beck, T. E. Kidd, C. C. Chancey, and Xuan P A Gao, Observation of infrared-active modes in Raman scattering from topological insulator nanoplates, Nanotechnology 23, 455703 (2012).

18. Gayatri D. Keskar, Ramakrishna Podila, Lihua Zhang, Apparao M. Rao, and Lisa D. Pfefferle, Synthesis and Raman Spectroscopy of Multiphasic Nanostructured Bi − Te Networks with Tailored Composition, The Journal of Physical Chemistry C 117, 9446 (2013).

19. M. K. Tran, J. Levallois, P. Lerch, J. Teyssier, A. B. Kuzmenko, G. Autès, O. V. Yazyev, A. Ubaldini, E. Giannini, D. van der Marel, and A. Akrap, Infrared- and Raman-spectroscopy measurements of a transition in the crystal structure and a closing of the energy gap of BiTeI under pressure, Physical Review Letters 112, 047402 (2014).

20. Chunxiao Wang, Xiegang Zhu, Louis Nilsson, Jing Wen, Guang Wang, Xinyan Shan, Qing Zhang, Shulin Zhang, Jinfeng Jia, and Qikun Xue, In situ Raman spectroscopy of topological insulator Bi_2Te_3 films with varying thickness, Nano Research 6, 688 (2013).

21. Ajay Soni, Zhao Yanyuan, Yu Ligen, Michael Khor Khiam Aik, Mildred S. Dresselhaus, and Qihua Xiong, Enhanced thermoelectric properties of solution grown $Bi_2Te_{3-x}Se_x$ Nanoplatelet Composites, Nano Letters 12, 1203 (2012).

22. Yujie Liang, Wenzhong Wang, Baoqing Zeng, Guling Zhang, Jing Huang, Jin Li, Te Li, Yangyang Song, and Xiuyu Zhang, Raman scattering investigation of Bi_2Te_3 hexagonal nanoplates prepared by a solvothermal process in the absence of NaOH, Journal of Alloys and Compounds 16, 5147 (2011).

23. C. Rodríguez-Fernández, C. V. Manzano, A. H. Romero, J. Martín, M. Martín-González, Morais de Lima M. Jr, and A. Cantarero, The fingerprint of Te-rich and stoichiometric Bi_2Te_3 nanowires by Raman spectroscopy, Nanotechnology 19, 075706 (2016).

24. V. Russo, A. Bailini, M. Zamboni, M. Passoni, C. Conti, C. S. Casari, A. Li Bassi, C. E. Bottani, Raman spectroscopy of Bi-Te thin films, Journal of Raman Spectroscopy 39, 205 (2008).

25. Yanyuan Zhao, Xin Luo, Jun Zhang, Junxiong Wu, Xuxu Bai, Meixiao Wang, Jinfeng Jia, Hailin Peng, Zhongfan Liu, Su Ying Quek, and Qihua Xiong, Interlayer vibrational modes in few-quintuple-layer Bi_2Te_3 and Bi_2Se_3 two-dimensional crystals: Raman spectroscopy and first-principles studies, Physical Review B 90, 245428 (2014).

Chapter 4 Electrical transport

4.1 Transport of Bi_2Se_3

4.1.1 Weak antilocalization and linear magnetoresistance

Topological insulators have symmetry-protected surface states within the bulk gap. As a result of the strong spin-orbit coupling (SOC), the spin-momentum locked surface states show weak antilocalization (WAL) effect due to strong spin-orbit coupling. Lu *et al.* proposed a magnetoconductivity formula for the surface states of a magnetically doped topological insulator, which reveals a competing effect of weak localization (WL) and weak antilocalization in quantum transport when an energy gap is opened at the Dirac point by magnetic doping.[1] They found that random magnetic scattering can drive the system from the symplectic to the unitary class, and the band gap can induce a crossover from weak antilocalization to weak localization. It is tunable by the Fermi energy. Figure 4.1 shows the band gap structure vs thickness.

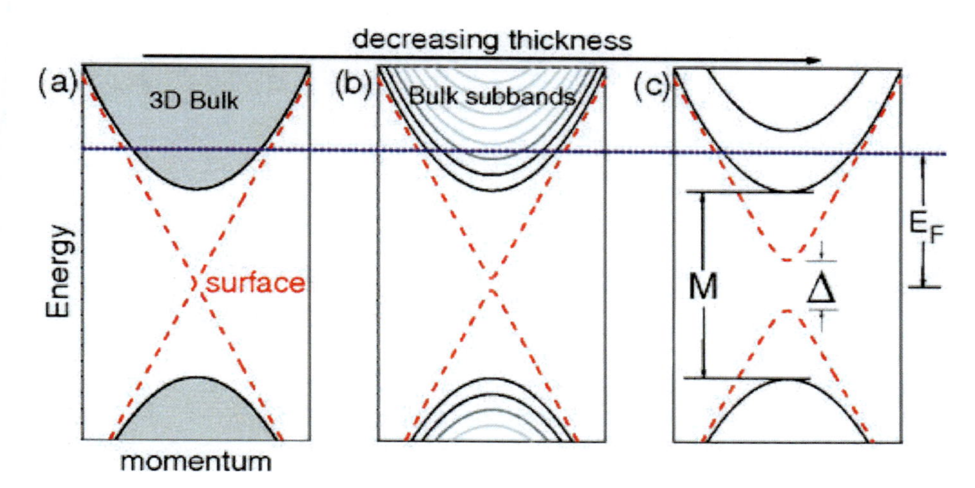

Figure 4.1 (a) The gapped bulk and gapless surface bands of a 3D topological insulator. (b) The quantum confinement along the z direction splits the 3D bulk bands into 2D subbands. The hybridization of the top and bottom surfaces opens a gap Δ for the gapless surface bands. (c) Decreasing the thickness to the ultrathin limit, the Fermi surface intersects with only one pair of bulk subbands (with band gap M) and one pair of gapped surface bands.[2]

Under different experimental conditions, the bulk states of topological insulators with strong SOC can lead to either WAL or WL. However, it is difficult to distinguish the bulk state features from the surface state behaviors by the perpendicular field transport measurements. WAL can be expected when a strong SOC or scattering comes into play. SOC in the bulk states is usually strong enough to lead to a topological phase transition.

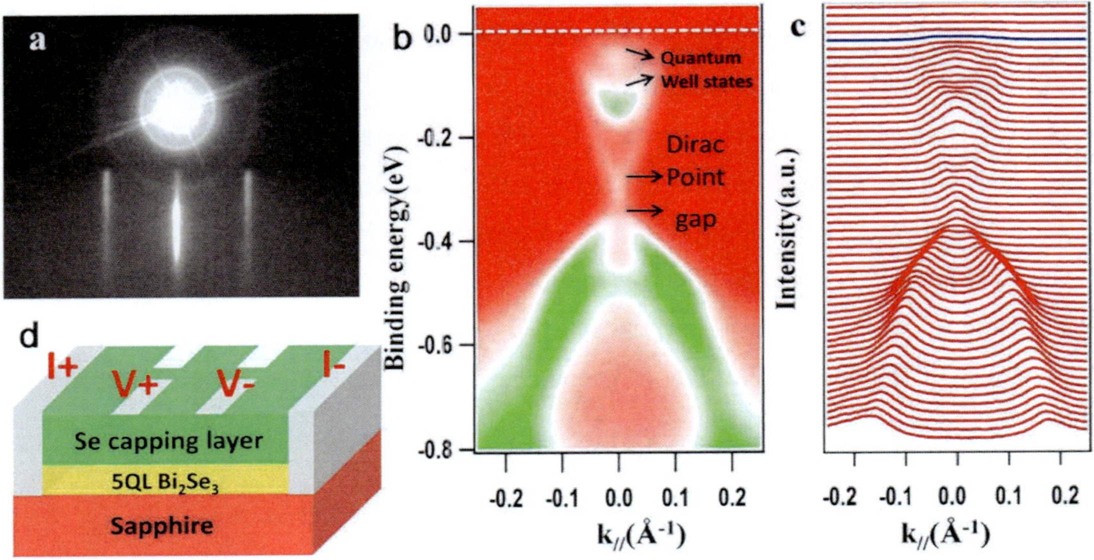

Figure 4.2 (a) Reflection high-energy electron diffraction (RHEED) pattern. (b) ARPES band spectra along the Γ-K direction. Fermi level is marked by the white dashed line. Quantum well states, Dirac point and small gap are marked by the arrows. (c) The corresponding momentum distribution curves of (b). Fermi level is marked by the blue line. (d) The schematic device structure for transport measurements.[3]

Wang *et al.* reported the 5 QLs Bi2Se3 films grown on sapphire (0001) substrates by MBE.[3] Figure 4.2(a) depicts the reflection high-energy electron diffraction pattern of Bi_2Se_3 films. Figure 4.2(b) shows the band map of the Bi_2Se_3 films taken in Γ–K direction. Moreover, Figures 4.2(b) and 4.2(c) show that the quantum well states of the conduction band contribute a lot to the Fermi surface, indicating the Dirac surface states and the bulk states playing an important role in the charge transport.

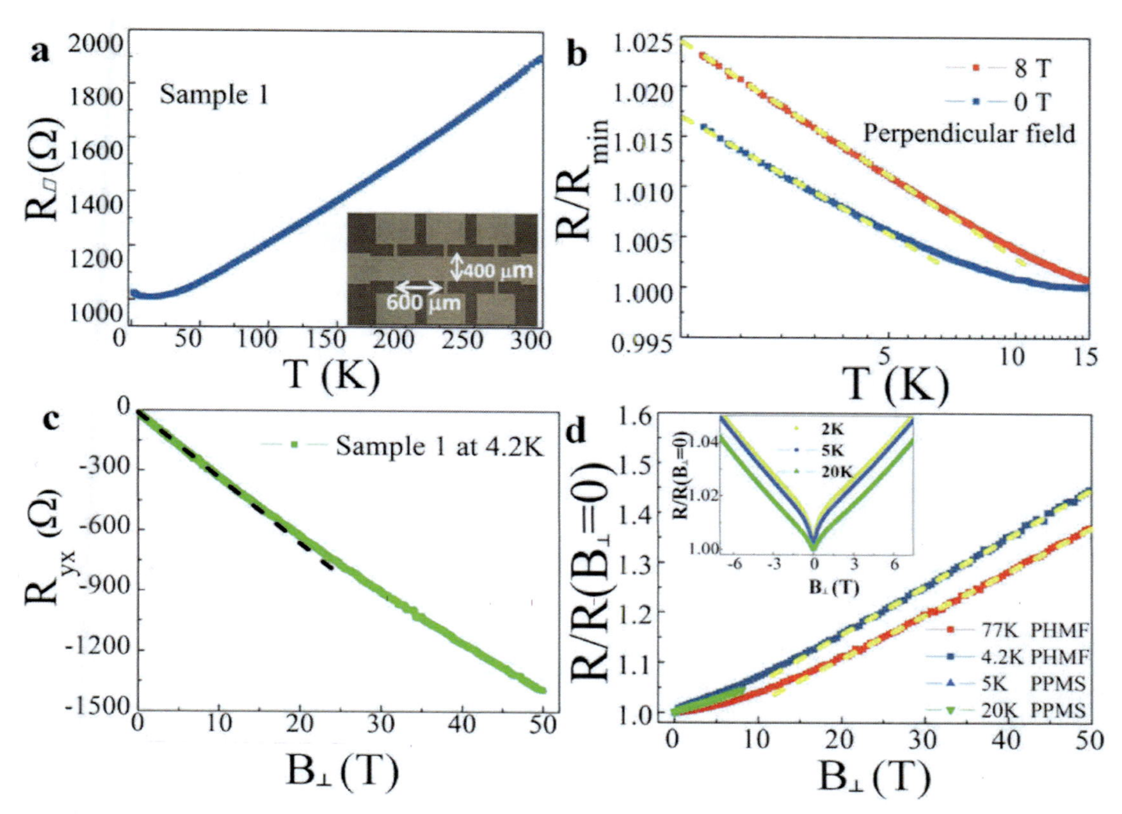

Figure 4.3 (a) The sheet resistance $R_\square$ as a function of temperature. Optical image of the Hall bar structure is shown in the inset. (b) Normalized upturn resistances at zero field and 8 T. (c) R_{yx} at 4.2 K as a function of magnetic field. (d) magneto-resistance (MR) properties of films. The inset shows the MR dips around 0 T due to the WAL effect at low temperatures. Dashed lines are guides to the eyes.[3]

The Hall bars with channel size of 0.6×0.4 mm^2 are fabricated in order to explore the transport properties of the films. Its optical image is shown in the inset of Figure 4.3(a). Temperature dependent sheet resistance $R_\square$ decreases nearly linearly with the decrease of temperature in the range between 300 K and 50 K, as shown in Figure 4.3(a). Figure 4.3(a) shows the resistance increases with the decrease of temperature below 12 K due to the quantum correction induced by the electron-electron interaction.[4] Such resistance increase becomes stronger in a larger magnetic field 8 T than 0 T, which can be explained by the suppression of the WAL effect. Figure 4.3(c) shows the R_{yx} at 4.2 K and Figure 4.3(d) shows dips of magnetic resistance around 0 T below 50 K due to the WAL effect. Linear magnetic resistance (LMR) can be observed above 16 T below 77 K. The origin of LMR, however, is interpreted by different ways, either a quantum origin or a classical origin. About quantum origin of LMR, the Abrikosov's quantum LMR model explain it as the linear energy dispersion of the gapless topological surface states,[5,6] while Hikami-Larkin-Nagaoka theory relates it to the WAL mechanism.[7-9] About classical origin of LMR, Parish and Littlewood interpreted the LMR to the sample inhomogeneity.[10]

For the TI films, LMR needs a very high magnetic field at low temperature (for example, 16 T and 77 K, as shown in Figure 4.3). Tang et al. reported LMR in Bi_2Se_3 nanoribbons with thickness from 50 to 400 nm and widths from 200 nm to several micrometer at a much lower magnetic field of 1-2 T even at room temperature.[5] This extends TI materials application for room temperature magneto-electronic device. One advantage of nanoribbon is its extremely high surface-to-volume ratio, which is useful to distinguish two dimensional surface transport from three dimensional bulk transport.

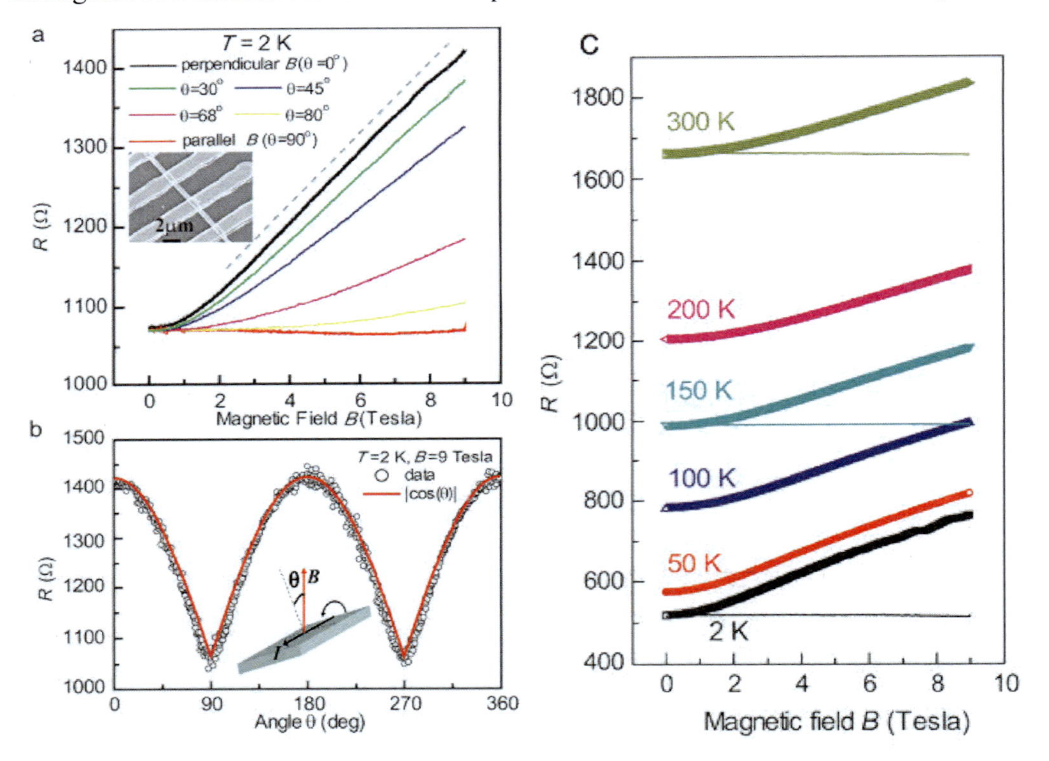

Figure 4.4 Two-dimensional LMR in Bi_2Se_3 nanoribbons at different tilt angle to the magnetic field. The gray dashed linear line is to guide the eye. The inset shows an SEM image of a nanoribbon. (b) Resistance as a function of tilt angle θ. The data can be fitted by $|cos(\theta)|$ (red line). (c) Resistance as a function of magnetic field below 300 K. The data in the perpendicular or parallel field are shown as dots or solid lines.[5]

Figure 4.4 shows that the largest magneto resistance is reached when the tilt angle of sample surface to the magnetic field is zero. LMR can be observed above 1 to 2 T, and even in room temperature. The angular dependence of the MR follows a $|\cos(\theta)|$ relationship, illustrating the two dimensional origin of LMR. At perpendicular magnetic field, the resistance showed weak oscillations on the top of LMR. These oscillations are Shubnikov-de Haas oscillations, which will be discussed in the following chapter.

4.1.2 Shubnikov-de Haas oscillations and Zeeman effect

At low temperature and in high magnetic field, an oscillation in the conductivity of a material can be observed. That is the so-called Shubnikov–de Haas effect (SdH), a macroscopic manifestation of the inherent quantum mechanical nature of matter. The analysis of SdH in Bi_2Se_3 remains unclear. Some researcher reported on single-band SdH oscillations due to bulk carriers or topological surface states.[11,12] Others, however, found multiband SdH oscillations and reported the difficulties to distinguish between the bulk and surface staes.[13,14] Veyrat *et al.* considered band bending effect in analysis of SdH effect in nanostructure of Bi_2Se_3.[15]

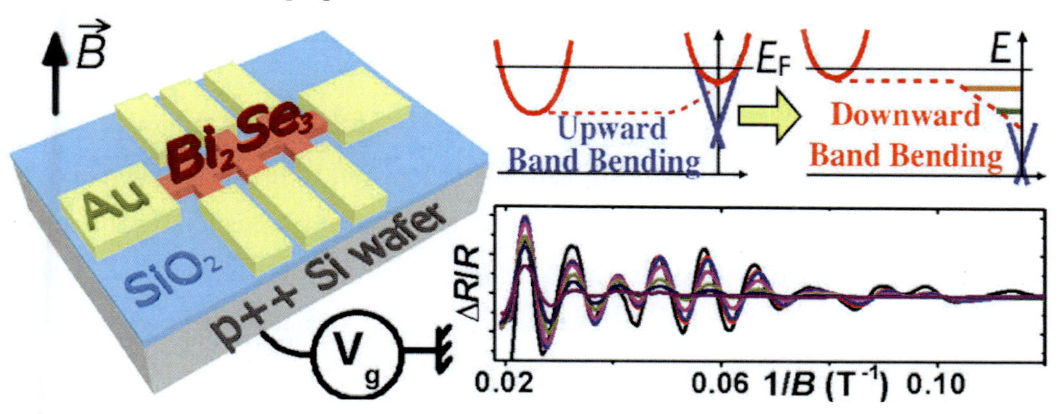

Figure 4.5 (a) Monocrystalline Bi_2Se_3 nanostructures grown by chemical vapor transport. Au contacts were prepared by e-beam lithography, metal lift-off and Ar-ion etching. Upward and downward band bendings are indicated. Shubnikov-de Haas oscillations were found under high magnetic field.[15]

Their investigated bulk carrier densities ranges from 3×10^{19} cm^{-3} to 6×10^{17} cm^{-3} and an upward to downward band bending can be observed at low bulk density sample due to the competition between bulk and interface doping.[15] The downward band bending is induced by the SiO_2 substrate, leading to a formation of a potential well at the interface in which a confined two dimensional electron gas coexists with topological surface states. To avoid downward band bending, controlling the interfaces quality is important. One way may be to grow the samples on oxygen-free substrates.

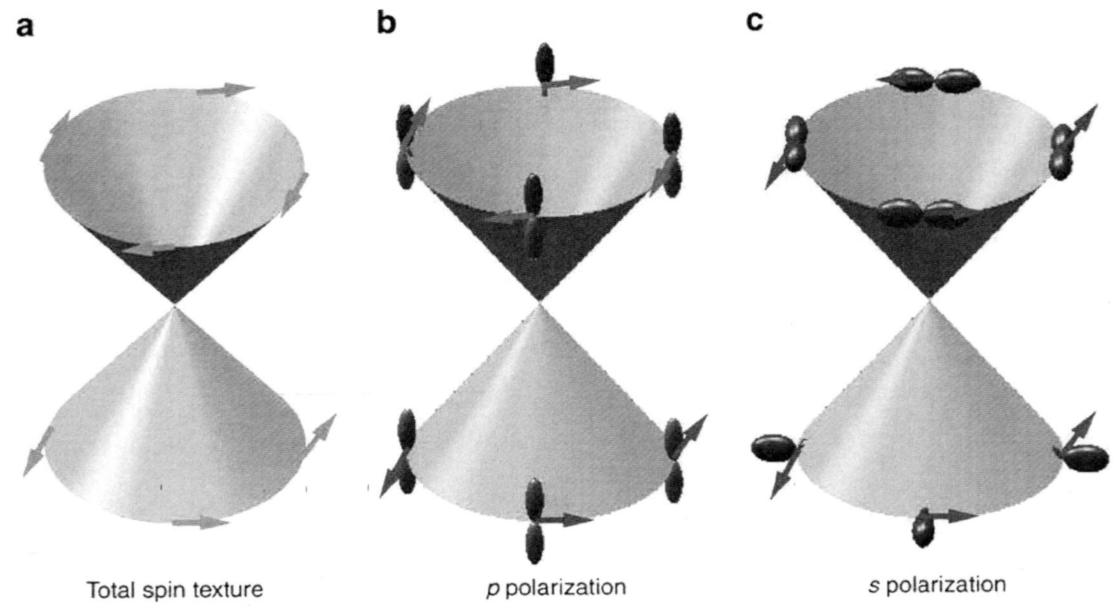

Figure 4.6 (a) Schematic of the total spin texture. The arrows show spin directions. Orbital textures probed in the (b) p-polarization and (c) s-polarization geometry and their related spin textures.[17]

One important characteristic of Bi_2Se_3 is the helical orbital texture of the surface states, a left-handed spin texture for the upper Dirac cone and a right-handed spin texture for the lower Dirac cone. For the upper Dirac cone, a left-handed spin texture is coupled to the radial orbital texture, right-handed to tangential. For the lower Dirac cone, the coupling is opposite. The tangential orbital texture is dominant for the upper Dirac cone with radial for the lower one. This results in the right-handed spin texture for the in-plane orbitals of both the upper and lower Dirac cones.[16] The different spin texture detected locking with the orbital texture in Bi_2Se_3 is the manifestation of the spin-orbital texture and the photoemission matrix element effects, as shown in Figure 4.6.[17] The surface states of Bi_2Se_3 have a large Landé factor $g= 50$ and helical spin–orbital texture, therefore, an in-plane magnetic field will induce Zeeman energy. This will lead to a deformation of the Dirac cone, which results in spin polarization of the surface states.

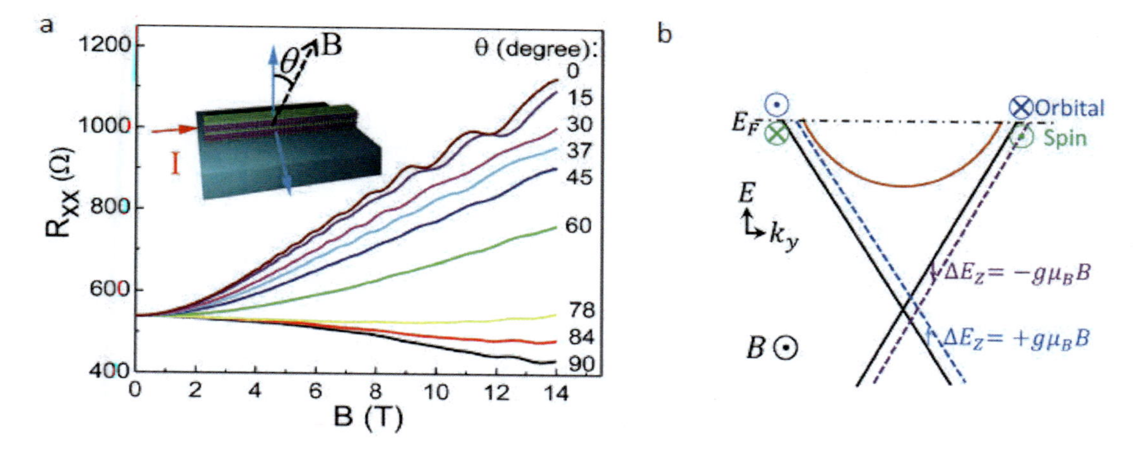

Figure 4.7 (a) Magneto-transport under a tilted magnetic field at the temperature of 2 K. The resistance R_{xx} vs magnetic field B at different angles. (b) In an in-plane magnetic field, the Zeeman energies, $+g\mu_B B$ and $-g\mu_B B$, lead to shifts of the surface energy band.[18]

Figure 4.7 shows magnetic field angle dependent magneto-transport. The evolution of the magneto resistance, measured at 2 K, as a function of angle from 0° to 90° is shown. Below a low magnetic field of 3 T, the magneto resistance obeys a quadratic dependence on magneto field, $R(B) \propto (\mu B\cos\theta)^2$, where μ is the mobility. SdH oscillations can be observed to shift to higher magnetic field with the increase of angle in high magnetic field. Magneto resistance turns from positive to negative for angle $\geq$78°. Such negative magneto resistance does not come from the weak localization, because it is very weak temperature dependence while for weak localization it is temperature dependent. The other reason is the weak localization induced negative magneto resistance will saturate around 1 T but in the experiment it does not saturated even at 14 T. The third reason is the magneto resistance exists at room temperature, in which a weak localization effect will not happen. The origin of the negative magneto resistance is due to Zeeman energy.[18] Zeeman energies, $+g\mu_B B$ and $-g\mu_B B$, add to the surface band in an in-plane magnetic field, where μ_B is the Bohr magneton. Zeeman energies go to the maximum in the parallel direction to the magnetic field and to the minimum (zero) in the perpendicular direction. Zeeman energies induce deformation of the surface Fermi circle. The rotational symmetry breaks the Fermi circle, leading to spin polarization.

4.1.3 Influence of thickness

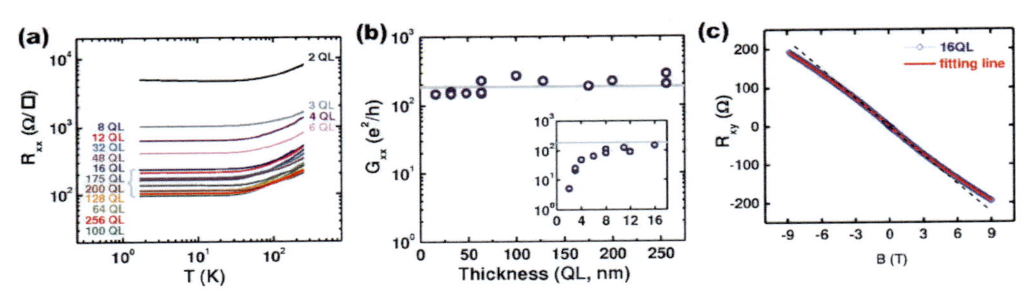

Figure 4.8 (a) Resistance Rxx(T), (b) conductance Gxx(t) at 1.5 K, and (c) Hall resistance Rxy(B) of a 16 QL.[19]

Unlike the thickness effect on Raman spectra of topological insulator Bi_2Se_3, Bansal *et al.* reported thickness independent transport properties in the range of 8 to 256 QL.[19] Figure 4.8(a) shows that the resistance for all the samples with the thicknesses from 2 to 256 QL decreases as the decrease of the temperature from 290 to 30 K and below 30 K the resistance almost does not change. As shown in Figure 4.8(b), the resistance/conductance is quite thickness independent in the range of 8 to 256 QL, indicating that the conductance in such thickness range is mainly due to some surface transport channels. This can be demonstrated by the nonlinear $R_{xy}(B)$, shown in Figure 4.8(c).

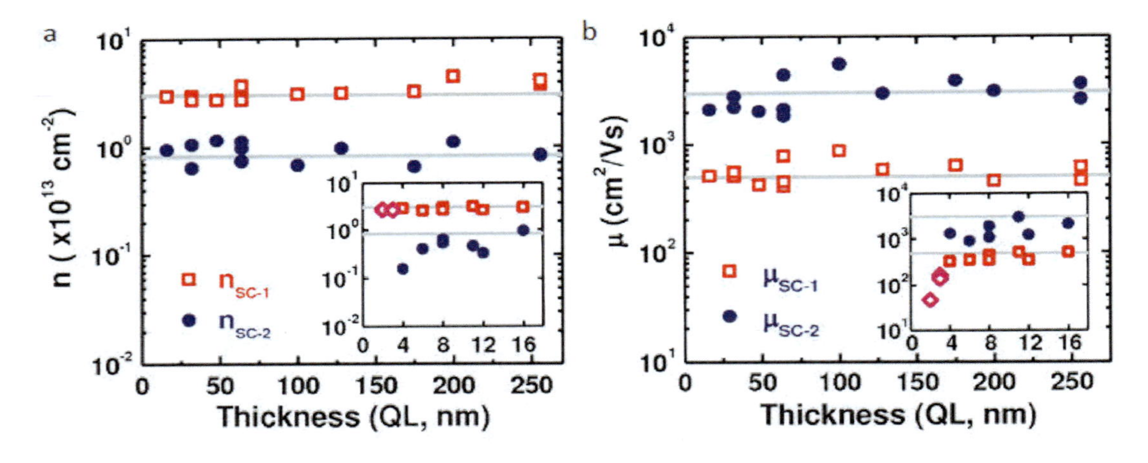

Figure 4.9 (a) Sheet carrier densities and (b) mobilities vs thickness.[19]

Given two carrier types dominate the Hall effect, the $R_{xy}(B)$ is related to two sheet carrier densities n_1 and n_2, and their respective mobility μ_1 and μ_2: $R_{xy}(B) = -\frac{B}{e}[n_1\mu_1^2 + n_2\mu_2^2 + B^2\mu_1^2\mu_2^2(n_1 + n_2)][(n_1\mu_1 + n_2\mu_2)^2 + B^2\mu_1^2\mu_2^2(n_1 + n_2)^2]^{-1}$, where e is the electron charge and B is the magnetic field. Figure 4.9 shows the sheet carrier densities n_1 and n_2 and mobilities μ_1 and μ_2 are nearly thickness independent. These two surface channels are consistent with the topological surface states and the surface accumulation layers.[19]

4.1.4 Influence of pressure

Topological insulator has a topologically protected surface state; however, it is difficult to identify it from the bulk conductivity in transport measurements. The study of pressure dependent topological insulator is important because on the one hand pressure can suppress the bulk conductivity and on the other hand pressure may drive topological insulator superconducting without destroying the surface state.

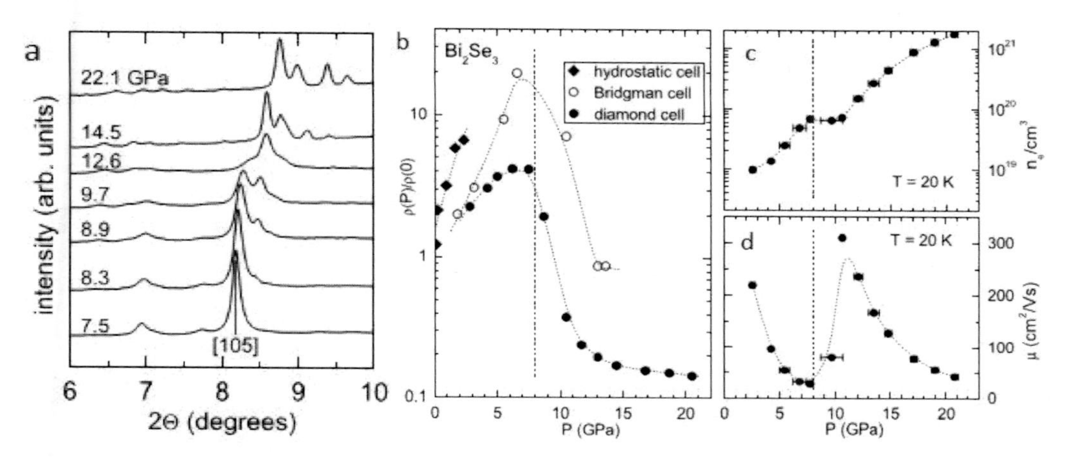

Figure 4.10 (a) Evolution of the x-ray diffraction pattern with pressure. A structural transition can be observed above 8 GPa. (b) Electrical resistivity as a function of pressure. The dashed line indicates the crystal structure change around 8 GPa. (c) Carrier density n and (d) mobility μ at 20 K versus pressure.[20]

The structure of Bi_2Se_3 remains rhombohedral phase below 8 GPa, while additional peaks can be observed above 8 GPa, as shown in Figure 4.10(a). Three cells experiments are used in the measurement, including hydrostatic, Bridgman, and diamond cells.[20] Their results are shown in Figure 4.10(b). The resistivity increases with the increase of pressure for all types of measurements. A structural transition is observed above 8 GPa, as indicated by an obvious drop in the resistance.

4.2 Doping effect of Bi_2Se_3

4.2.1 Ca doping

The bulk chemical potential of as grown Bi_2Se_3 is in the conduction band. Proper chemical doping can result in chemical potential falling from the conduction band into the gap and then into the valence band. Ca doped Bi_2Se_3 shows a p-type behavior through low level substitutions.[21]

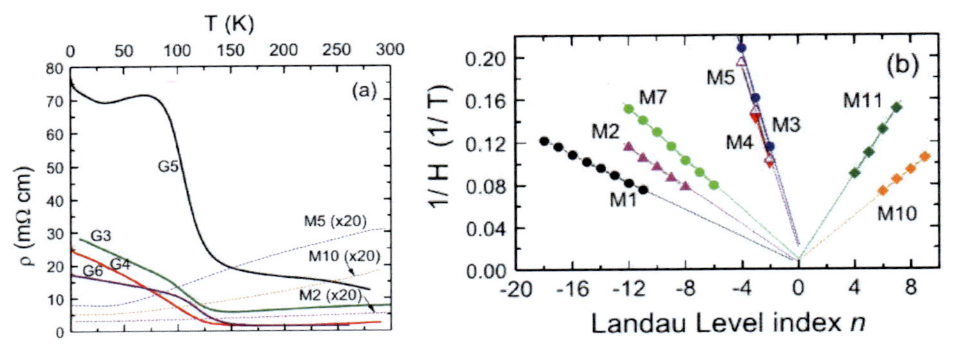

Figure 4.11 (a) Resistivity ρ *of Ca doped* Bi_2Se_3 *as a function of* temperature *T. Resistivity* ρof *samples M2, M5, and M10 show a metallic temperature dependence while those of samples G3, G4, G5, and G6 display insulating behavior. (b) Landau Level index n as a function of magnetic field minima in the SdH oscillations in metallic samples from M1 to M11. Negative Landau Level index n represents the electron Fermi surface and positive one is the hole Fermi surface.[22]*

Resistivity ρ of samples M2, M5, and M10 increase with increase of the temperature, exhibiting a metallic behavior, as shown in Figure 4.11. For samples G3, G4, and G6, resistivity ρ shows opposite trend, as it decrease up heating, and the values are dozens of mΩ below 100 K, display insulating behaviors. Landau Level index *n* in Figure 4.11(b) displays that caliper area S_F of the bulk Fermi surface decreases for samples M1 to M3, Se vacancies pin μ_b goes into the gap from the conduction band. Then it increases as Se vacancies pin μ_b moves from the gap into the valence band.

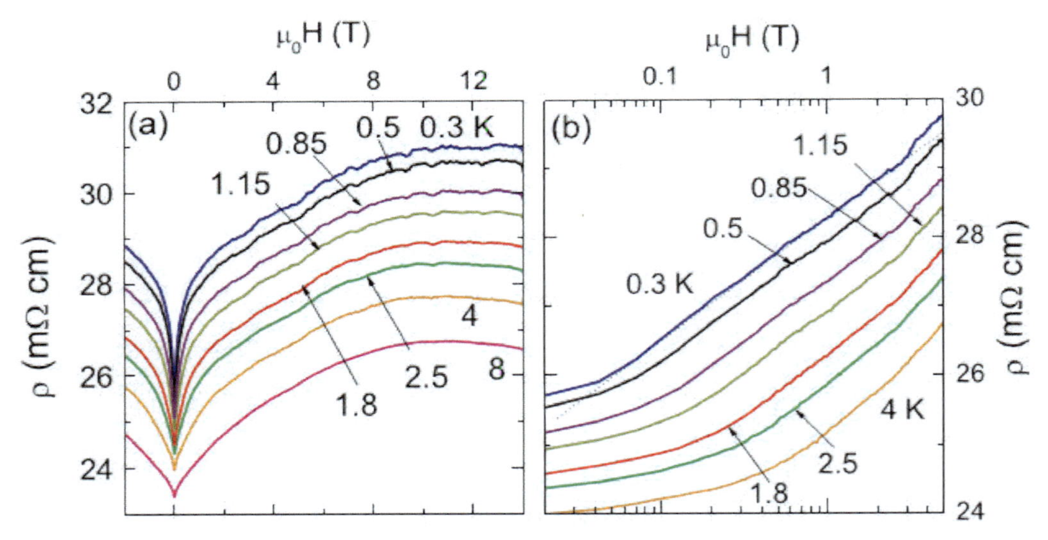

Figure 4.12 Resistivity ρ *of sample G4 as a function of magnetic field in the temperature range of 0.3 and 8 K in (c) linear and (d) log scale, respectively.[22]*

Resistivity as a function of magnetic field in the temperature range below 4 K is shown in Figure 4.12. Obvious weak field anomaly can be observed, with a deep valley at zero field. A logarithmic behavior can be seen near the valley at 0.3 K. The results of nonmetallic samples are related to long phase-breaking lengths in a conductance channel. The origin of the in-gap conducting states is associated to the topological surface states.

4.2.2 Sm doping

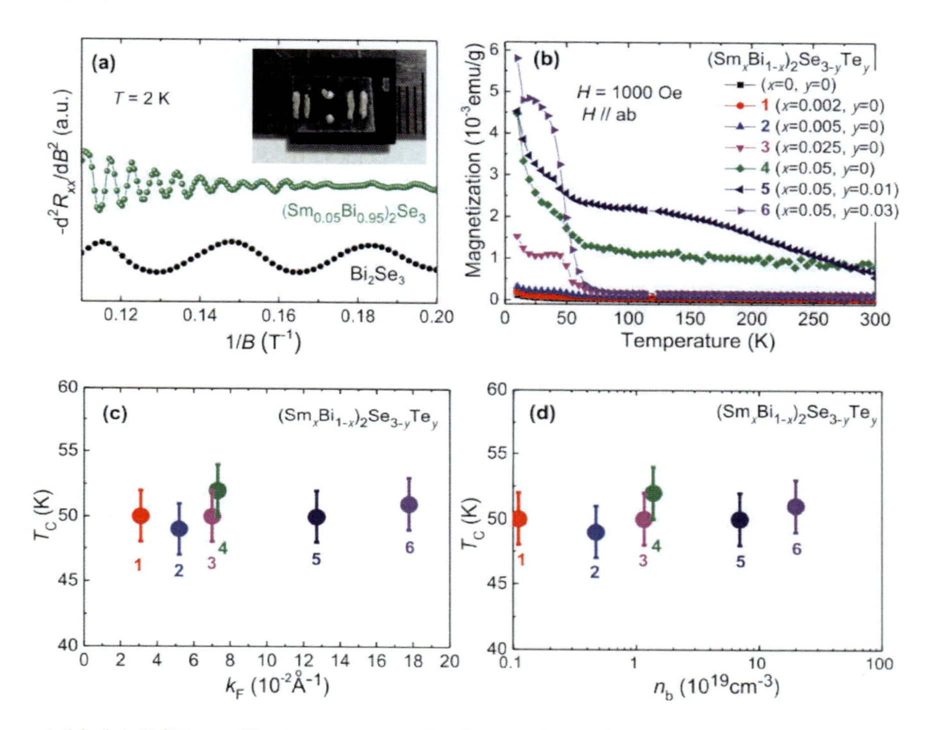

Figure 4.13 (a) SdH oscillations in pure Bi_2Se_3 and Sm doped Bi_2Se_3 crystals. A Hall bar device is shown in the inset. (b) Magnetization as a function of temperature. Curie temperature T_C as a function of (c) Fermi wave vector k_F (d) and carrier concentration n_b.[29]

Magnetically doped topological insulators are a new type of dilute magnetic semiconductors, attracting great interest in the topological magneto electric effect and the quantum spin/anomalous Hall effect.[23,24] Surface electrons of three-dimensional topological insulators are helical Dirac fermions protected from backscattering by time-reversal symmetry in the case of no magnetic doping. After doping with magnetic ions in topological insulator, it is ferromagnetic, and time-reversal symmetry is broken. Magnetic topological insulators expand the limits of magnetic characterization capability. To realize novel quantum states and devices based on topological insulators, scientists have tried to induce ferromagnetic states through either doping or interfacial proximity effects.[25-27] Lee *et al.* reported structural, magnetic, and magneto transport properties of Mn-doped Bi_2Te_3. They found that the crystal structure changes from the tetradymite structure to a BiTe phase with increasing Mn concentration. Ferromagnetism starts with a Curie temperature in the range 13.8–17 K in films in the range of 2%–10% Mn concentration.[28]

Sm-doped Bi_2Se_3 is a candidate high-mobility topological insulator with ferromagnetism up to 52 K and a very low bulk electron carrier concentration of 10^{18} cm^{-3}.[29] Transport and magnetic properties of samples Bi_2Se_3, Sm doped Bi_2Se_3 and $(Sm_xBi_{1-x})_2Se_{3-y}Te_y$ are shown in Figure 4.13. Introduction of Te atoms results in additional *n*-type carriers, and is expected to tune the carrier concentration as well as the Fermi wave vector k_F. The Fermi wave vector k_F can be extracted from the SdH oscillation while the carrier concentration n_b can be extracted from Hall effect measurements. Curie temperature T_C does not show an obvious variation of the Fermi wave vector k_F and the carrier concentration n_b over an order, indicating a robust ferromagnetism.

4.2.3 Cr doping

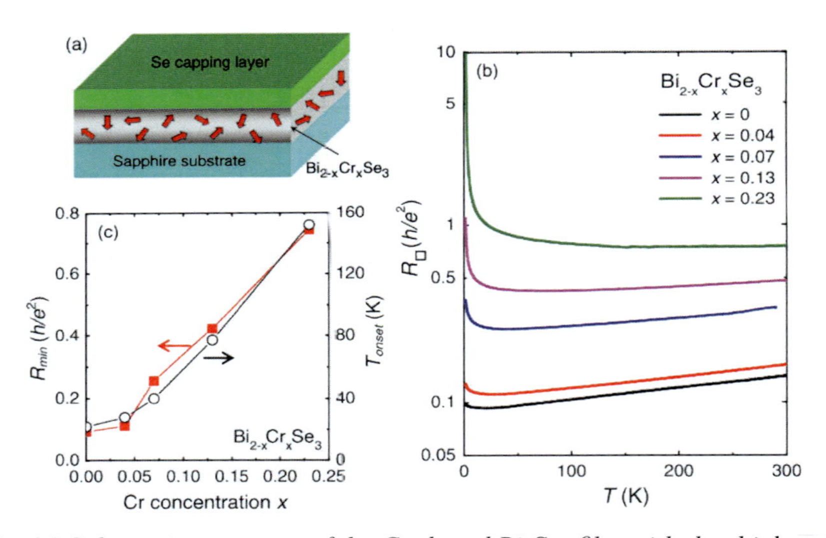

Figure 4.14 (a) Schematic structure of the Cr doped Bi_2Se_3 film with the thickness of 3 QL. (b) Sheet resistance of Cr doped Bi_2Se_3 film as a function of temperature. (c) The minimum resistance and the onset temperature as a function of Cr concentration.[9]

The interest in study of magnetic doping of Bi_2Se_3 has been growing fast due to the fact that Bi_2Se_3 is an ideal 3D topological insulator with a large bulk energy gap (300 meV). An opening of the energy gap at the Dirac point can be observed in the three-dimensional topological insulator dibismuth triselenide with magnetic dopants.[8] Liu *et al.* reported magneto conductivity change from weak antilocalization to weak localization with the increase of Chromium (Cr) concentration.[9] The ground state is dominated by weak localization in the high Chormium concentration area. Upon heating weak antilocalization appears while weak localization comes back again under strong magnetic field. Such crossover phenomenon reveals a phase transition from topological insulator to a dilute magnetic semiconductor due to the doping magnetic impurities effect. Figure 4.14 (a) shows the 3 QL thick films, where surface state dominates in transport measurement due to a high surface to volume ratio. Figure 4.14 (b) shows the sheet resistance as a function of temperature for Cr doped Bi_2Se_3 films, which exhibit metallic behavior at high temperature and insulating behavior at low temperature (roughly 30 K or less). Cr doping leads to enhance the insulating tendency, suppressing the electrical conductivity. The minimum resistance increases 1 order from roughly 0.1 to 1 with the increase of Cr concentration. The onset temperature has the similar trend, as it increase from about 20 K to 150 K. At rather low temperature below 30 K, the resistance of the highest Cr doping sample at 23% is two orders of magnitude larger than that of pure sample. The increase of minimum resistance and onset temperature as a function of Cr concentration reflects an insulating behavior and a weak Kondo effect.

Reference

1. Hai-Zhou Lu, Junren Shi, and Shun-Qing Shen, Competition between weak localization and antilocalization in topological surface states, Physical Review Letters 107, 076801 (2011).

2. Hai-Zhou Lu and Shun-Qing Shen, Weak localization of bulk channels in topological insulator thin films, Physical Review B 84, 125138 (2011).

3. Huichao Wang, Haiwen Liu, Cui-Zu Chang, Huakun Zuo, Yanfei Zhao, Yi Sun, Zhengcai Xia, Ke He, Xucun Ma, X. C. Xie, Qi-Kun Xue, and Jian Wang, Crossover between weak antilocalization and weak localization of bulk states in ultrathin Bi_2Se_3 films, Scientific Reports 4, 5817 (2014).

4. Jian Wang, Ashley M. DaSilva, Cui-Zu Chang, Ke He, J. K. Jain, Nitin Samarth, Xu-Cun Ma, Qi-Kun Xue, and Moses H. W. Chan, Evidence for electron-electron interaction in topological insulator thin films, Physical Review B 83, 245438 (2011).

5. Hao Tang, Dong Liang, Richard L. J. Qiu, and Xuan P. A. Gao, Two-dimensional transportinduced linear magneto-resistance in topological insulator Bi_2Se_3 nanoribbons, ACS Nano 5, 7510 (2011).

6. Xiaolin Wang, Yi Du, Shixue Dou, and Chao Zhang, Room Temperature Giant and Linear Magnetoresistance in Topological Insulator Bi_2Te_3 Nanosheets, Physical Review Letters 108, 266806 (2012).

7. Shinobu Hikami Anatoly I. Larkin Yosuke Nagaoka, Y. Spin-orbit interaction and magnetoresistnce in the two dimensional random system, Progress of Theoretical Physics 63, 707 (1980).

8. S. X. Zhang, R. D. McDonald, A. Shekhter, Z. X. Bi, Y. Li, Q. X. Jia, and S. T. Picraux, Magneto-resistance up to 60 Tesla in topological insulator Bi_2Te_3 thin films, Applied Physics Letters 101, 202403 (2012).

9. B. A. Assaf, T. Cardinal, P. Wei, F. Katmis, J. S. Moodera, and D. Heiman, Linear magnetoresistance in topological insulator thin films: Quantum phase coherence effects at high temperatures, Applied Physics Letters 102, 012102 (2013).

10. Jifa Tian, Cuizu Chang, Helin Cao, Ke He, Xucun Ma, Qikun Xue, and Yong P. Chen, Quantum and classical magnetoresistance in ambipolar topological insulator transistors with gate-tunable bulk and surface conduction, Scientific Reports 4, 4859 (2014).

11. James G. Analytis, Jiun-Haw Chu, Yulin Chen, Felipe Corredor, Ross D. McDonald, Z. X. Shen, and Ian R. Fisher, Bulk Fermi surface coexistence with Dirac surface state in Bi_2Se_3: A comparison of photoemission and Shubnikov–de Haas measurements, Physical Review B 81, 205407 (2010).

12. Yuan Yan, Li-Xian Wang, Xiaoxing Ke, Gustaaf Van Tendeloo, Xiao-Song Wu, Da-Peng Yu, and Zhi-Min Liaoa, High-mobility Bi_2Se_3 nanoplates manifesting quantum oscillations of surface states in the sidewalls, Scientific Reports 4, 3817 (2014).

13. Benjamin Sacépé, Jeroen B. Oostinga, Jian Li, Alberto Ubaldini, Nuno J.G. Couto, Enrico Giannini, and Alberto F. Morpurgo, Gate-tuned normal and superconducting transport at the surface of a topological insulator, Nature Communications 2, 575 (2011).

14. Cheng Zhang, Xiang Yuan, Kai Wang, Zhi-Gang Chen, Baobao Cao, Weiyi Wang, Yanwen Liu, Jin Zou and Faxian Xiu, Observations of a metal–insulator transition and strong surface states in $Bi_{2-x}Sb_xSe_3$ thin films, Advanced Materials 26, 7110 (2014).

15. Louis Veyrat, Fabrice Iacovella, Joseph Dufouleur, Christian Nowka, Hannes Funke, Ming Yang, Walter Escoffier, Michel Goiran, Barbara Eichler, Oliver G. Schmidt, Bernd Büchner, Silke Hampel, and Romain Giraud, Band Bending Inversion in Bi_2Se_3 Nanostructures, Nano Letters 15, 7503 (2015).

16. Haijun Zhang, Chao-Xing Liu, and Shou-Cheng Zhang, Spin-orbital texture in topological insulators, Physical Review Letters 111, 066801 (2013).

17. Zhuojin Xie, Shaolong He, Chaoyu Chen, Ya Feng, Hemian Yi, Aiji Liang, Lin Zhao, Daixiang Mou, Junfeng He, Yingying Peng, Xu Liu, Yan Liu, Guodong Liu, Xiaoli Dong, Li Yu, Jun Zhang, Shenjin Zhang, Zhimin Wang, Fengfeng Zhang, Feng Yang, Qinjun Peng, Xiaoyang Wang, Chuangtian Chen, Zuyan Xu, and X. J. Zhou, Orbital-selective spin texture and its manipulation in a topological insulator, Nature Communications 5, 3382 (2014).

18. Li-Xian Wang, Yuan Yan, Liang Zhang, Zhi-Min Liao, Han-Chun Wu, and Da-Peng Yu, Zeeman effect on surface electron transport in topological insulator Bi_2Se_3 nanoribbons, Nanoscale 7, 16687 (2015).

19. Namrata Bansal, Yong Seung Kim, Matthew Brahlek, Eliav Edrey, and Seongshik Oh, Thickness-independent transport channels in topological insulator Bi_2Se_3 thin films, Physical Review Letters 109, 116804 (2012).

20. J. J. Hamlin, J. R. Jeffries, N. P. Butch, P. Syers, D. A. Zocco, S. T. Weir, Y. K. Vohra, J. Paglione, and M. B. Maple, High pressure transport properties of the topological insulator Bi_2Se_3, Journal of Physics : Condensed Matter 24, 035602 (2012).

21. Y. S. Hor, A. Richardella, P. Roushan, Y. Xia, J. G. Checkelsky, A. Yazdani, M. Z. Hasan, N. P. Ong, and R. J. Cava, P-type for topological insulator and low-temperature thermoelectric applications, Physical Review B 79, 195208 (2009).

22. J. G. Checkelsky, Y. S. Hor, M.-H. Liu, D.-X. Qu, R. J. Cava, and N. P. Ong, Quantum interference in macroscopic crystals of nonmetallic Bi_2Se_3, Physical Review Letters 103, 246601 (2009).

23. Xufeng Kou, Shih-Ting Guo, Yabin Fan, Lei Pan, Murong Lang, Ying Jiang, Qiming Shao, Tianxiao Nie, Koichi Murata, Jianshi Tang, Yong Wang, Liang He, Ting-Kuo Lee, Wei-Li Lee, Kang L. Wang, Scale-invariant dissipationless chiral transport in magnetic topological insulators beyond the two-dimensional limit, Physical Review Letters 113, 137201 (2014).

24. Abhinav Kandala, Anthony Richardella, Susan Kempinger, Chao-Xing Liu, and Nitin Samarth, Giant anisotropic magnetoresistance in a quantum anomalous Hall insulator, Nature Communications 6, 7434 (2015).

25. Cui-Zu Chang, Jinsong Zhang, Xiao Feng, Jie Shen, Zuocheng Zhang, Minghua Guo, Kang Li, Yunbo Ou, Pang Wei, Li-Li Wang, Zhong-Qing Ji, Yang Feng, Shuaihua Ji, Xi Chen, Jinfeng Jia, Xi Dai, Zhong Fang, Shou-Cheng Zhang, Ke He,

26. Yayu Wang, Li Lu, Xu-Cun Ma, and Qi-Kun Xue, Experimental observation of the quantum anomalous Hall effect in a magnetic topological insulator, Science 340, 167 (2013).

27. K. Carva, J. Kudrnovský, F. Máca, V. Drchal, I. Turek, P. Baláž, V. Tkáč, V. Holý, V. Sechovský, and J. Honolka, Electronic and transport properties of the Mn-doped topological insulator Bi2Te3: A first-principles study, Physical Review B 93, 214409 (2016).

28. A. J. Grutter, Perspective: Probing 2-D magnetic structures in a 3-D world, APL Materials 4, 032402 (2016).

29. Joon Sue Lee, Anthony Richardella, David W. Rench, Robert D. Fraleigh, Thomas C. Flanagan, Julie A. Borchers, Jing Tao, and Nitin Samarth, Ferromagnetism and spin-dependent transport in n-type Mn-doped bismuth telluride thin films, Physical Review B 89, 174425 (2014).

30. Taishi Chen,Wenqing Liu, Fubao Zheng, Ming Gao, Xingchen Pan, Gerrit van der Laan, Xuefeng Wang, Qinfang Zhang, Fengqi Song, Baigeng Wang, Baolin Wang, Yongbing Xu, Guanghou Wang, and Rong Zhang, High-mobility Sm-doped Bi2Se3 ferromagnetic topological Insulators and robust exchange coupling, Advanced Materials 27, 4823 (2015).

31. Y. L. Chen, J.-H. Chu, J. G. Analytis, Z. K. Liu, K. Igarashi, H.-H. Kuo, X. L. Qi, S. K. Mo, R. G. Moore, D. H. Lu, M. Hashimoto, T. Sasagawa, S. C. Zhang, I. R. Fisher, Z. Hussain, and Z. X. Shen, Massive Dirac fermion on the surface of a magnetically doped topological insulator, Science 329, 5992 (2010).

32. Minhao Liu, Jinsong Zhang, Cui-Zu Chang, Zuocheng Zhang, Xiao Feng, Kang Li, Ke He, Li-li Wang, Xi Chen, Xi Dai, Zhong Fang, Qi-Kun Xue, Xucun Ma, and Yayu Wang, Crossover between weak antilocalization and weak localization in a magnetically doped topological insulator, Physical Review Letters 108, 036805 (2012).

Chapter 5 Infrared spectroscopy

5.1 Optical conductivity

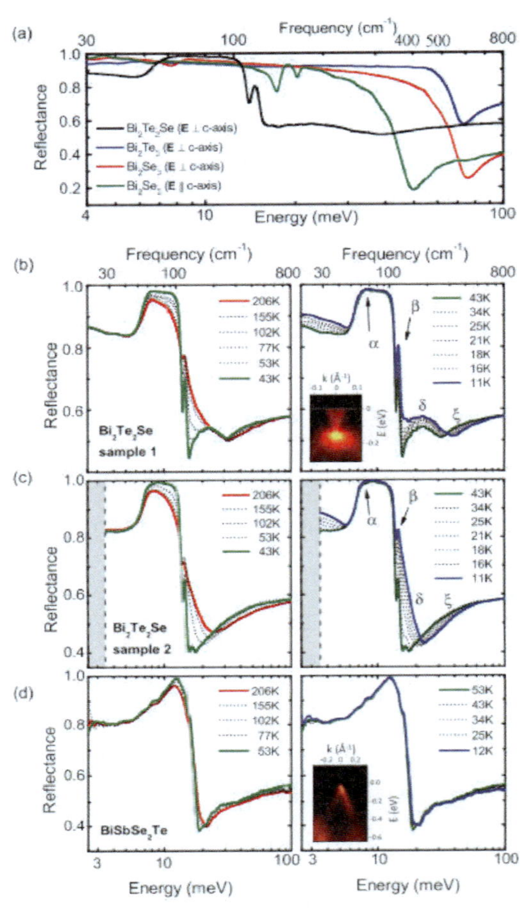

Figure 5.1 (a) Infrared reflectance of topological insulators Bi_2Se_3, Bi_2Te_3, and Bi_2Te_2Se from 4 to 100 meV, corresponding to 30 to 800 cm^{-1}. Plasma frequency can be seen at around 500 cm^{-1} for Bi_2Se_3 and Bi_2Te_3. (b) and (c) Reflectance of topological insulators Bi_2Te_2Se from 11 K to 206 K. Inset shows the ARPES. Four spectral features α, β, δ, and ξ are marked. The optical properties of sample 1 and sample 2 are very similar. (d) Reflectance of topological insulators Bi_2SbSe_2Te from 12 K to 206 K. Inset shows the ARPES.[4]

As discussed above, at low temperatures in transport experiments, topological insulator surface state conductance can be observed, while at room temperature surface conductance is strongly suppressed in transport experiments. [1-3] However, ARPES experiments still can resolve the surface states at room temperature. So the interesting question is why the surface state conductance is suppressed in transport experiments at room temperature. Luckily, temperature dependence of optical spectroscopy is a useful method for elucidating the transport suppression mechanism and surface-to-bulk coupling in topological insulators, because such measurements can provide the information about carrier scattering rates, effective mass, optical bandgap, phonons, and their respective coupling.

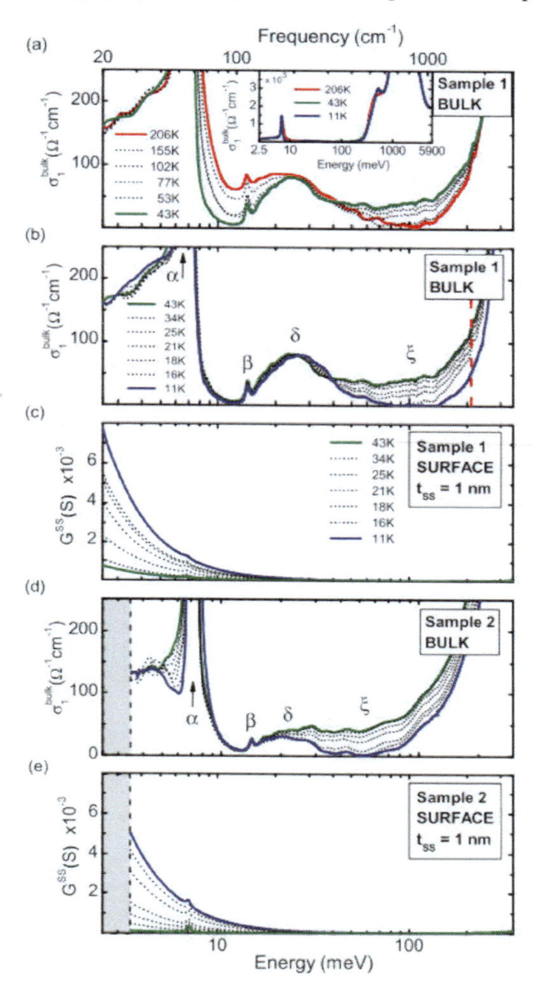

Figure 5.2 (a)(b)Temperature dependent real part of the bulk optical conductivity of Bi_2Te_2Se. (c) Temperature dependent real part of the surface conductance. Below 43 K, a Drude mode and coherent surface transport can be observed. Inset shows the bulk optical conductivity in a wide range. The real part of (d) the bulk optical conductivity and (e) the surface conductance of the Bi_2Te_2Se.[4]

A plasma frequency can be seen at around 75 meV, as shown in Figure 5.1 (a).[4] Reflectance spectra of Bi_2Te_2Se and Bi_2SbSe_2Te are vey different as shown in Figure 5.1(b)(c)(d). For Bi_2Te_2Se, it contains a peak followed by a plateau from 60 to 100 cm^{-1} and a depression due to an optical phonon. The real part of the bulk optical conductivity and surface conductance are shown in Figure 5.2. It is dominated by α phonon mode at around 7 meV and β phonon mode at around 14 meV. With decreasing the temperature, α phonon mode sharpens. Both α phonon mode and β phonon mode are E_u symmetry, indicating strong electron-phonon coupling and quantum interference between the phonon and an electronic continuum, best described by a Fano line shape.[5]

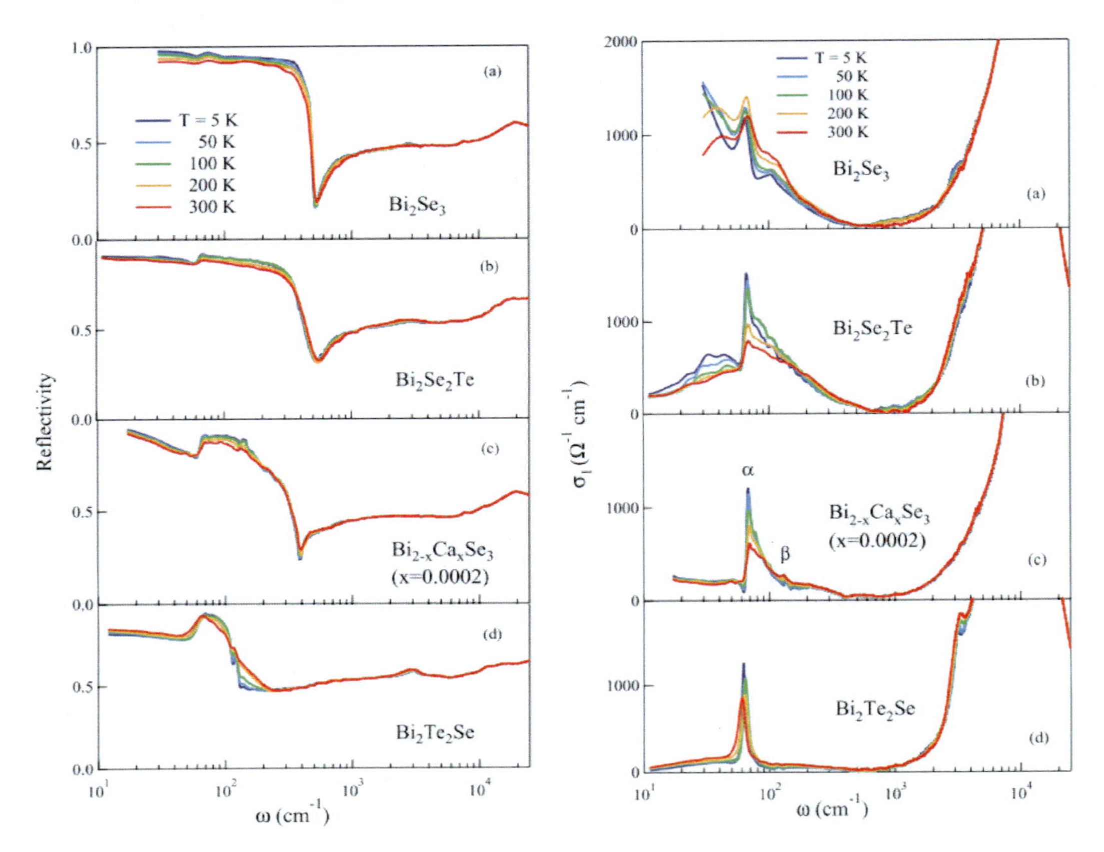

Figure 5.3 A wide range of temperature dependent (Left) reflectivity and (Right) real part of the optical conductivity of (a) Bi_2Se_3, (b) Bi_2Se_2Te, (c) $Bi_{1.9998}Ca_{0.0002}Se_3$, and (d) Bi_2SeTe_2. A free-carrier plasma edge can be observed around 400 to 500 cm^{-1}. Two phonon modes α and β are at about 60 and 130 cm^{-1}. A small bump at around 3000 cm^{-1} is due to the direct-gap transition.[6]

With decrease of the temperature, the Drude term in Bi_2Se_3 superimposed to the two phonon peaks, become sharper, as shown in Figure 5.3.[6] For Bi_2Se_2Te, the effect of compensation becomes observable. For $Bi_{1.9998}Ca_{0.0002}Se_3$ and Bi_2SeTe_2, the spectral weight reduces. Note that most spectral weight in the far-infrared region is located at finite frequencies in the phonon region.

5.2 Plasmon and charge inhomogeneity

Terahertz Kerr measurements on topological insulator suggested a bulk charge non-uniform distribution, therefore, it is interesting to detect inhomogeneities by a contactless and nondestructive infrared spectroscopy.[7] Dordevic et al. reported a plasma edge in Bi_2Se_3, Bi_2Te_3 and Sb_2Te_3 and found their different temperature dependence.[8]

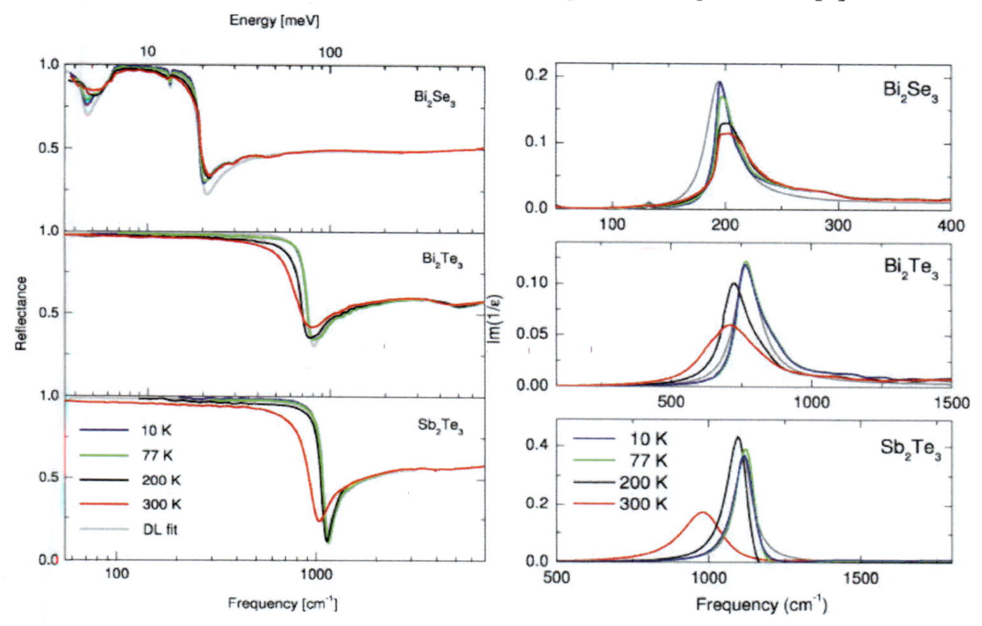

Figure 5.4 Reflectance spectra and loss function of Bi_2Se_3, Bi_2Te_3, and Sb_2Te_3. The plasma edge of these three topological insulators becomes sharper with decrease of the temperature. Note that with decrease of the temperature the plasma minimum is shifting to lower frequencies in Bi_2Se_3, to higher frequencies in Bi_2Te_3 and Sb_2Te_3. Two strong phonon modes (α and β, as discussed above) are observed in Bi_2Se_3, but not observed in Bi_2Te_3 and Sb_2Te_3. The big asymmetry in the shape of the loss function peak may be related to the presence of charge inhomogeneities.[8]

Temperature dependence of infrared spectra for Bi_2Se_3, Bi_2Te_3, and Sb_2Te_3 are shown in Figure 5.4, with a well defined plasma edge. The plasma frequency ω_p^* is related to the carrier density n by $\omega_p^{*2} = \frac{4\pi n e^2}{m^* \varepsilon_\infty}$, where m^* is the carrier effective mass and ε_∞ is the high-frequency dielectric function. As we known, plasmon is a longitudinal mode and the real part of the optical conductivity does not couple to it. The loss function are shown in Fgiure 5.4. The loss function reveals the presence of longitudinal modes, including plasmons. Obvious asymmetry, that is deviation from the expected Lorentzian-like shape of the plasmon, can be observed in the loss function, indiating electronic inhomogeneities.

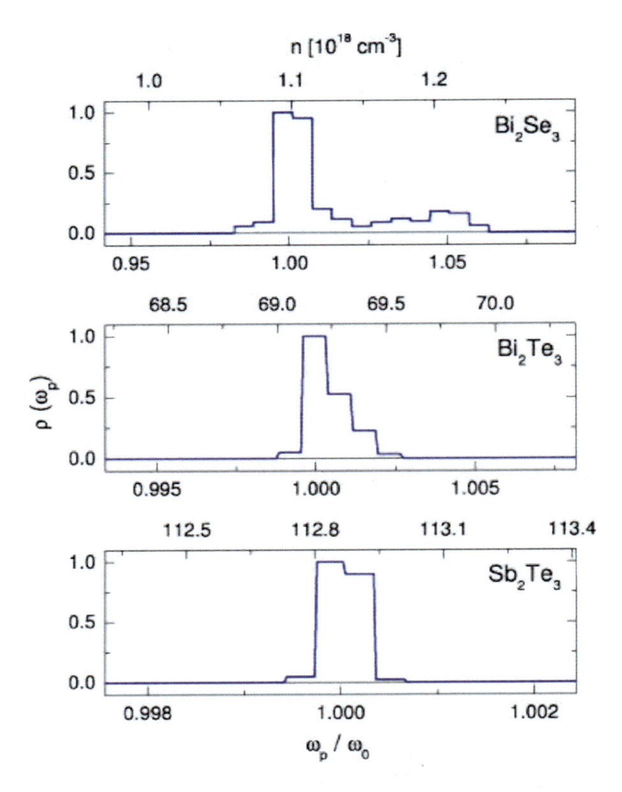

Figure 5.5 Histogram representation of the distribution function of plasma frequencies $\rho(\omega_p)$ vs. normalized plasma frequency ω_p/ω_0 (bottom axes) and carrier density n (top axes) for Bi_2Se_3, Bi_2Te_3, and Sb_2Te_3. The data is collected at 10 K and all the distribution functions are normalized.[8]

The quantitative information about the asymmetry, can be fitted by an equation proposed by Van der Marel and Tsvetkov.[9] The expression is $\frac{1}{\epsilon(\omega)} = \int_0^{\infty} \frac{1}{\epsilon_i(\omega,\omega_p)} \boldsymbol{\rho(\omega_p)d\omega_p}$, where $\boldsymbol{\epsilon(\omega)}$ is the complex dielectric function, $\boldsymbol{\rho(\omega_p)}$ is the distribution function of plasma frequencies, and $\boldsymbol{\epsilon_i(\omega, \omega_p)}$ is the complex dielectric function for each individual component. Based on this expression and histogram method, the distribution function of plasma frequencies $\boldsymbol{\rho(\omega_p)}$ is shown in Figure 5.5. The distribution of Sb_2Te_3 is narrow and nearly symmetric, while it becomes broader and slightly asymmetric in Bi_2Te_3 and more broader and asymmetry in Bi_2Se_3 due to different type of charge carriers, that is holes compared to electrons.

5.3 Thickness dependent electronic properties

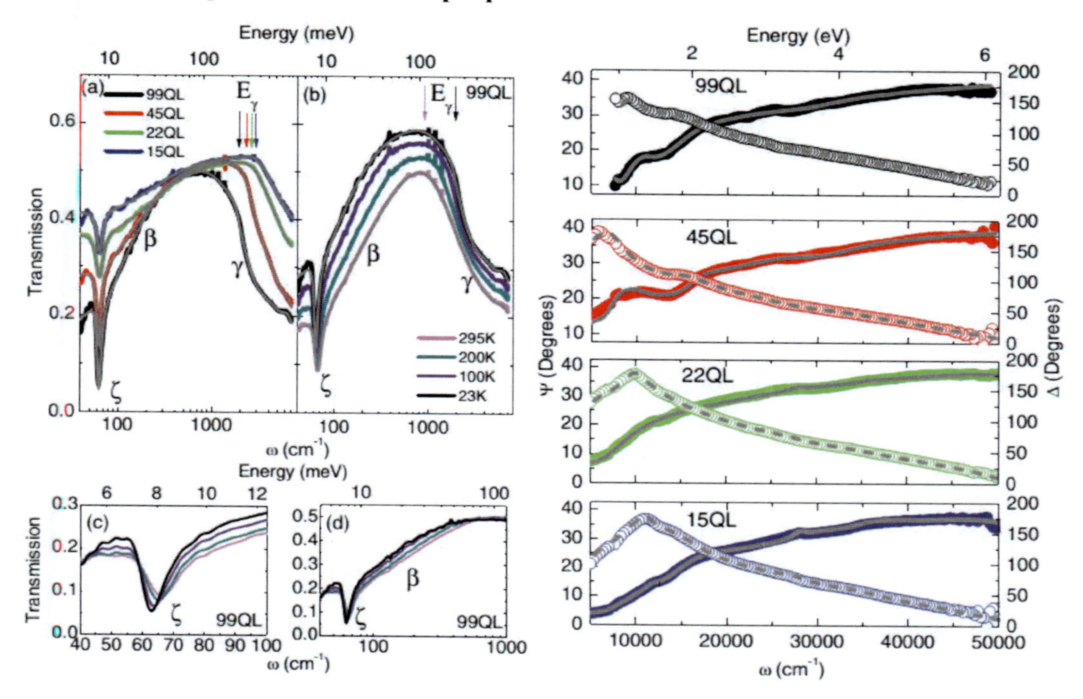

Figure 5.6 (a) Transmittance spectra of Bi_2Se_3 films for thicknesses in the range from 15 to 99 QL at 20 K. (b) Transmittance spectra of 99 QL Bi_2Se_3 film from 20 K to 295 K. The spectra are offset by 0.03 for clarity. E_γ indicates the optical gap. (c) (d)Temperature dependent infrared transmittance spectra of 99 QL film. (Right) The experimental and fitting ellipsometry for all films are at 20 K. The open circles are the measured value of (left axis) while the solid circles are the measured values of (right axis). The gray line is the fitting value.[10]

Among all known TIs, Bi_2Se_3 has the largest bandgap with a topologically protected surface states band structure and a simple Dirac cone with the Dirac point lying directly above the valence band. In order to know the thickness influence on the electronic structure and dynamics, Post et al. investigated Bi_2Se_3 films with thicknesses 15, 22, 45, and 99 QL using a combination of using variable angle spectroscopic elipsometry and transmission for the range 5 meV to 6 eV (40 cm^{-1} to 50000 cm^{-1}).[10] Figure 5.6 shows three main features ζ, β, and γ, corresponding to an infrared active phonon, free carriers, and energy gap. As we known, transmission is exponentially related to the thickness. The thicker sample is, the lower transmission is expected.

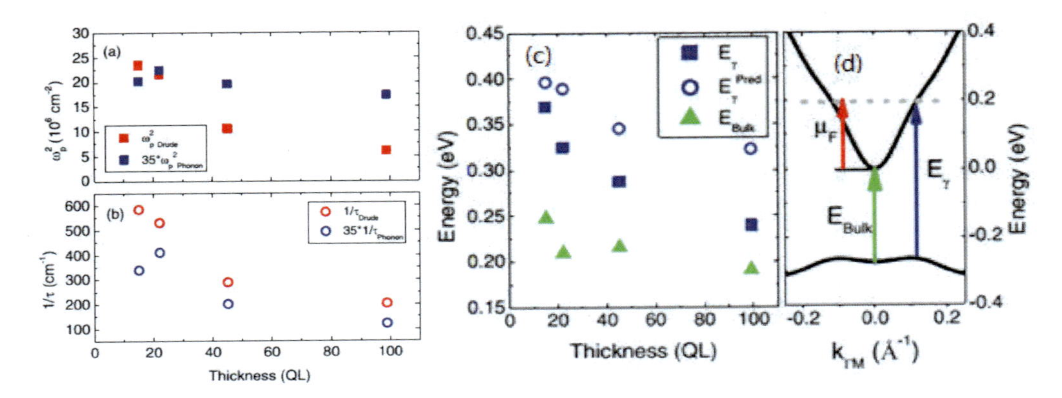

Figure 5.7 (a) The experimental oscillator strength for the Drude peak in red and the phonon. (b)1/ꞇ is the scattering rate for the Drude and phonon.(c) The value of E_γ determined from the fit, E_γ^{Pred} determined by combining optical data with photoemission studies and the bulk energy gap E_{bulk}. All the energy gap decreases with increasing thickness.[10]

The oscillator strength of the free carriers $\omega_{p,Drude}^2$ and the phonon $\omega_{p,phonon}^2$ are shown in Figure 5.7. In order to compare the magnitude of $\omega_{p,Drude}^2$, $\omega_{p,phonon}^2$ is multiplied by 35. $\omega_{p,phonon}^2$ does not show obvious thickness dependent, while $\omega_{p,Drude}^2$ decreases with increase of thickness. Scattering rate *1/ꞇ* of Drude and phonon increases as thickness decreases, indicating the free carriers being from impurity doping. It is strongest in thinner samples. This means if there are more carriers, there are more defects, leading to a higher scattering rate for both the phonon and the free carriers. The bulk energy gap is smaller in these thin films than most reported value 300 meV due to in-gap impurity states.

5.4 Magneto optical spectroscopy of $Bi_{2-x}Sb_xTe_{3-y}Se_y$

Dirac-like surface states are widely studied in topological insulators. Isolating the surface states charge carriers from the bulk response is not always so easy. Actually, it has been only reported in a few cases.[12-14] The surface states are always weak, masked by bulk dopants because vacancies and antisite defects influence the chemical potential in the bulk bands.[15] One common way to overcome this problem is to combine n-type Bi_2Se_3/Bi_2Te_3 with the p-type Sb_2Te_3, that is $Bi_{2-x}Sb_xTe_{3-y}Se_y$ (BSTS). In this case, p-type and n-type defects compensate each other, leading to the bulk insulating. Figure 5.8 show the energy gap of BSTS1 and BSTS2. For BSTS1, the linear region in $(\varepsilon_2)^2$ shows shift to higher energy with the decrease of the temperature above 100 K. Below 100 K, it shifts to lower energy and a second linear region appears, related to a second interband transition. For

BSTS2 in the measured temperature region, only one linear region can be observed and it monotonically increases upon cooling. The band gap is 270 meV at 300 K and it increases to 350 meV at 10 K.

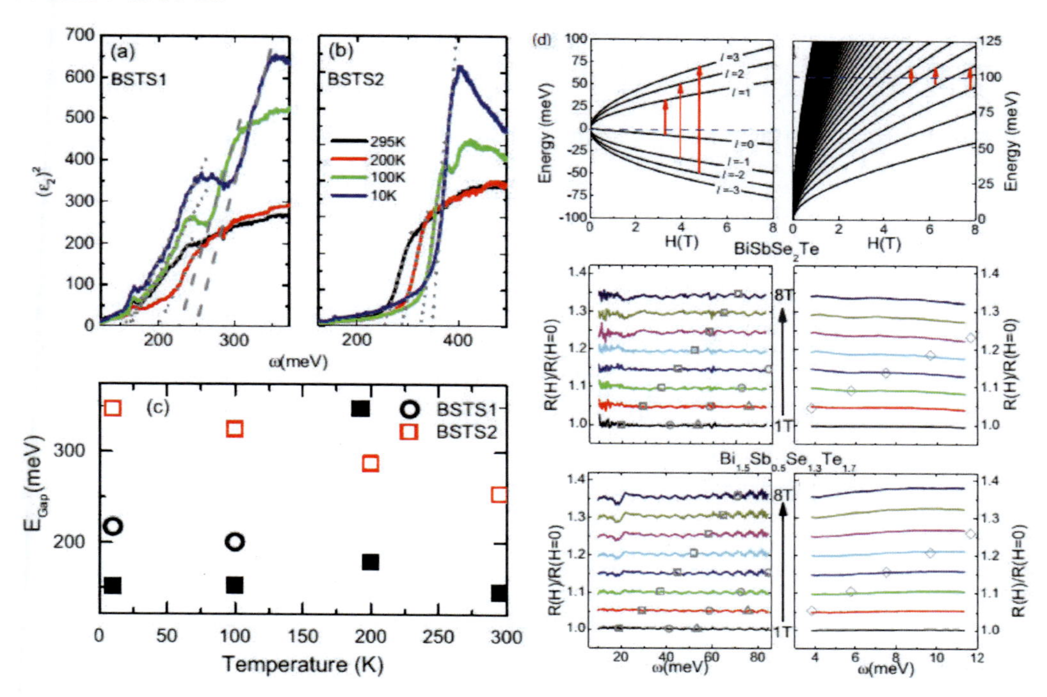

Figure 5.8 (a) Temperature dependent $(\varepsilon_2)^2$ spectra of $BiSbTe_2Se$ (BSTS1) and $Bi_{1.5}Sb_{0.5}Te_{1.3}Se_{1.7}$ (BSTS2) from 10 to 295 K. The direct band gap is indicated by the linear region of the linear region of $(\varepsilon_2)^2$. (c) The values of E_{Gap} are plotted. A second linear region appears in BSTS1 below 100 K, plotted as the black open circles.(d) The Landau levels dispersion of the surface states. Red arrows show the allowed transitions and dashed blue lines show Fermi energy E_F at the Dirac point. Normalized reflectance is plotted, with increasing fields offset by 0.05.Experimental data are shown by squares, circles and triangles, indicating the expected positions of inter Landau levels. Intra Landau levels transitions are shown in the red arrows, when E_F is far above the Dirac point.[11]

Magneto-optical spectroscopy allows the measurements of the wavelength dependence of the magnetic field induced changes of polarization parameters and rotation of the light reflected or transmitted by magnetic samples. The principle of magneto-optical spectroscopy is related to the fact that electrons undergo cyclotron motion in the crossed electric and magnetic fields at the frequency of $\omega_c=eB/m^*$.[16] The cyclotron motion leads to the splitting of the energy bands into discrete Landau levels, which can be distinguished by their index l. These Landau levels disperse with field depending on the effective mass of the charge carriers.[17,18]

Infrared magneto-optical spectra of BSTS are shown in Figure 5.8 (d). For BSTS1, E_F is at the Dirac point. The allowed inter Landau levels transitions are from l_{-n} to l_{n+1}. The magnetic field does not affect the spectra too much. For BSTS2, E_F is above the Dirac point. Therefore, lots of inter Landau levels transitions are not allowed. Like to the BSTS1, neither intra nor inter Landau levels can be observed in BSTS2. This can be explained by the strongly disordered surface states than the bulk states and small surface carrier density.

5.5 Bandgap of Bi-based topological insulators

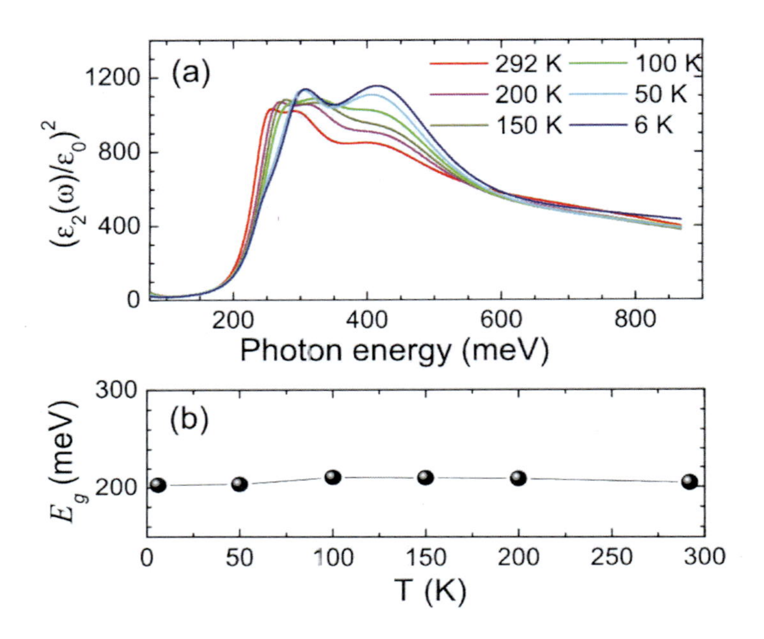

Figure 5.9 (a) Temperature dependent $(\varepsilon_2/\varepsilon_0)^2$ spectra of $(Bi,Sb)_2Te_3$) from 6 to 295 K. The band gap Eg can be extracted from the linear fit. (b) The values of E_g are plotted as a function of temperature.[21]

Topological insulator Bi_2Te_3 films are usually n-type bulk conductors,[19] while Sb_2Te_3 films are p-type bulk conductors.[20] One way is to grow $(Bi,Sb)_2Te_3$ to produce a compensated material with bulk insulating properties. Figure 5.9 shows temperature dependent band gap of $(Bi,Sb)_2Te_3$, determined by the linear part of $(\varepsilon_2)^2$, where ε_2 is the imaginary part of the dielectric function.[21] The band gap does not show obvious temperature dependence, as its value is around 200 meV below 300 K. The band gap of Bi-based topological insulators are usually several hundreds meV, listed in the table 5.1.

Bi-based TI material	band gap at room temperature	Ref.
Bi_2Se_3	300	[22]
Bi_2Te_3	142	[19]
$(Bi,Sb)_2Te_3$	207	[21]
$Bi_{1.5}Sb_{0.5}Te_{1.3}Se_{1.7}$	270	[11]
$(Bi_{0.9933}Sn_{0.0067})_2Te_3$	165	[23]
$[PbBi_2Te_2S_2]_{5SL}$	340	[24]
$[Bi_2Te_2S]_{1QL}$	300	[24]
$[Bi_2Te_2S]_{6QL}$	270	[24]
$[BiSbTe_2S]_{1QL}$	390	[24]

Table 5.1 Band gap of some Bi-based topological insulators. SL and QL represent septuple and quintuple layer, respectively.

The field of topological insulators is developing very fast, with new materials found often. For example, Antimonene oxide, a kind of novel 2D topological insulator with bandgap of 177 meV is reported very recently, which can be tunable in a wide range, useful for future flexible electronics and optoelectronics devices.[25]

Reference

1. Zhi Ren, A. A. Taskin, Satoshi Sasaki, Kouji Segawa, and Yoichi Ando, Large bulk resistivity and surface quantum oscillations in the topological insulator Bi_2Te_2Se, Physical Review B 82, 241306(R) (2010).

2. Hadar Steinberg, Dillon R. Gardner, Young S. Lee, and Pablo Jarillo-Herrero, Surface state transport and ambipolar electric field effect in Bi_2Se_3 nanodevices, Nano Letters 10, 5032 (2010).

3. Dohun Kim, Qiuzi Li, Paul Syers, Nicholas P. Butch, Johnpierre Paglione, S. Das Sarma, and Michael S. Fuhrer, Intrinsic electron-phonon resistivity of Bi_2Se_3 in the topological regime, Physical Review Letters 109, 166801 (2012).

4. Anjan A. Reijnders, Y. Tian, L. J. Sandilands, G. Pohl, I. D. Kivlichan, S. Y. Frank Zhao, S. Jia, M. E. Charles, R. J. Cava, Nasser Alidoust, Suyang Xu, Madhab Neupane, M. Zahid Hasan, X. Wang, S. W. Cheong, and K. S. Burch, Optical evidence of surface state suppression in Bi-based topological insulators, Physical Review B 89, 075138 (2014).

5. U. Fano, Effects of configuration interaction on intensities and phase shifts, Physical Review 124, 1866 (1961).

6. P. Di Pietro, F. M. Vitucci, D. Nicoletti, L. Baldassarre, P. Calvani, R. Cava, Y. S. Hor, U. Schade, and S. Lupi, Optical conductivity of bismuth-based topological insulators, Physical Review B 86, 045439 (2012).

7. G. S. Jenkins, A. B. Sushkov, D. C. Schmadel, N. P. Butch, P. Syers, J. Paglione, and H. D. Drew, Terahertz Kerr and reflectivity measurements on the topological insulator Bi_2Se_3, Physical Review B 82, 125120 (2010).

8. S. V. Dordevic, M. S. Wolf, N. Stojilovic, Hechang Lei and C. Petrovic, Signatures of charge inhomogeneities in the infrared spectra of topological insulators Bi_2Se_3, Bi_2Te_3 and Sb_2Te_3, Journal of Physics: Condensed Matter 25, 075501 (2013).

9. D. van der Marel and A. Tsvetkov, Transverse optical plasmons in layered superconductors, Czechoslovak Journal of Physics 46, 3165 (1996).

10. K. W. Post, B. C. Chapler, Liang He, Xufeng Kou, Kang L. Wang, and D. N. Basov, Thickness-dependent bulk electronic properties in Bi_2Se_3 thin films revealed by infrared spectroscopy, Physical Review B 88, 075121 (2013).

11. K. W. Post, Y. S. Lee, B. C. Chapler, A. A. Schafgans, Mario Novak, A. A. Taskin, Kouji Segawa, M. D. Goldflam, H. T. Stinson, Yoichi Ando, and D. N. Basov, Infrared probe of the bulk insulating response in $Bi_{2-x}Sb_xTe_{3-y}Se_y$ topological insulator alloys, Physical Review B 91, 165202 (2015).

12. A. A. Taskin, Zhi Ren, Satoshi Sasaki, Kouji Segawa, and Yoichi Ando, Observation of Dirac holes and electrons in a topological insulator, Physical Review Letters 107, 016801 (2011).

13. Dohun Kim, Sungjae Cho, Nicholas P. Butch, Paul Syers, Kevin Kirshenbaum, Shaffique Adam, Johnpierre Paglione, and Michael S. Fuhrer, Nature Physics 8, 459 (2012).

14. S. Y. Hamh, S.-H. Park, S.-K. Jerng, J. H. Jeon, S. H. Chun, J. H. Jeon, S. J. Kahng, K. Yu, E. J. Choi, S. Kim, S.-H. Choi, N. Bansal, S. Oh, Joonbum Park, Byung-Woo Kho, Jun Sung Kim, and J. S. Lee, Surface and interface states of Bi_2Se_3 thin films investigated by optical second-harmonic generation and terahertz emission, Applied Physics Letters 108, 051609 (2016).

15. J. Horák, Z. Stary, P. Lošák, J. Pancíř, Anti-site defects in n-Bi_2Se_3 crystals, Journal of Physics and Chemistry of Solids 51, 1353 (1990).

16. E. D. Palik and J. Furdyna, Handbook of the physics of thin-film solar cells, Reports on Progress in Physics 33, 1193 (1970).

17. A. A. Schafgans, S. J. Moon, B. C. Pursley, A. D. LaForge, M. M. Qazilbash, A. S. Sefat, D. Mandrus, K. Haule, G. Kotliar, and D. N. Basov, Electronic correlations and unconventional spectral weight transfer in the high-temperature pnictide superconductor using infrared spectroscopy, Physical Review Letters 108, 147002 (2012).

18. Chao-Xing Liu, Xiao-Liang Qi, HaiJun Zhang, Xi Dai, Zhong Fang, and Shou-Cheng Zhang, Model Hamiltonian for topological insulators, Physical Review B 82, 045122 (2010).

19. B. C. Chapler, K. W. Post, A. R. Richardella, J. S. Lee, J. Tao, N. Samarth, and D. N. Basov, Infrared electrodynamics and ferromagnetism in the topological semiconductors Bi_2Te_3 and Mn-doped Bi_2Te_3. Physical Review B 89, 235308 (2014).

20. Yeping Jiang, Y. Y. Sun, Mu Chen, Yilin Wang, Zhi Li, Canli Song, Ke He, Lili Wang, Xi Chen, Qi-Kun Xue, Xucun Ma, and S. B. Zhang, Fermi-level tuning of epitaxial Sb_2Te_3 thin films on graphene by regulating intrinsic defects and substrate transfer doping, Physical Review Letters 108, 066809 (2012).

21. K. W. Post, B. C. Chapler, M. K. Liu, J. S. Wu, H. T. Stinson, M. D. Goldflam, A. R. Richardella, J. S. Lee, A. A. Reijnders, K. S. Burch, M. M. Fogler, N. Samarth, and D. N. Basov, Sum-rule constraints on the surface state conductance of topological insulators, Physical Review Letters 115, 116804 (2015).

22. L. Wu, M. Brahlek, R. Valdés Aguilar, A. V. Stier, C. M. Morris, Y. Lubashevsky, L. S. Bilbro, N. Bansal, S. Oh, and N. P. Armitage, A sudden collapse in the transport lifetime across the topological phase transition in $(Bi_{1-x}In_x)_2Se_3$, Nature Physics 9, 410 (2013).

23. Y. L. Chen, J. G. Analytis, J. H. Chu, Z. K. Liu, S. K. Mo, X. L. Qi, H. J. Zhang, D. H. Lu, X. Dai, Z. Fang, S. C. Zhang, I. R. Fisher, Z. Hussain, Z. X. Shen, Experimental realization of a three-dimensional topological insulator Bi_2Te_3, Science 325, 178 (2009).

24. T. V. Menshchikova, M. M. Otrokov, S. S. Tsirkin, D. A. Samorokov, V. V. Bebneva, A. Ernst, V. M. Kuznetsov, and E. V. Chulkov, Band structure engineering in topological insulator based heterostructures, Nano Letters 13, 6064 (2013).

25. Shengli Zhang, Wenhan Zhou, Yandong Ma, Jianping Ji, Bo Cai, Shengyuan A. Yang, Zhen Zhu, Zhongfang Chen, and Haibo Zeng, Antimonene oxides: emerging tunable direct bandgap semiconductor and novel topological insulator, Nano Letters DOI: 10.1021/acs.nanolett.7b00297 (May, 2017).